Springer Proceedings in Complexity

Springer Proceedings in Complexity publishes proceedings from scholarly meetings on all topics relating to the interdisciplinary studies of complex systems science. Springer welcomes book ideas from authors. The series is indexed in Scopus.

Proposals must include the following:

- name, place and date of the scientific meeting
- a link to the committees (local organization, international advisors etc.)
- scientific description of the meeting
- list of invited/plenary speakers
- an estimate of the planned proceedings book parameters (number of pages/articles, requested number of bulk copies, submission deadline)

Submit your proposals to: Hisako.Niko@springer.com

More information about this series at https://link.springer.com/bookseries/11637

Zining Yang · Elizabeth von Briesen
Editors

Proceedings of the 2021 Conference of The Computational Social Science Society of the Americas

Springer

Editors
Zining Yang
Technology and Operations Management
California State Polytechnic University
Pomona, CA, USA

Elizabeth von Briesen
Assistant Professor of Computer Science
Elon University
Elon, NC, USA

ISSN 2213-8684 ISSN 2213-8692 (electronic)
Springer Proceedings in Complexity
ISBN 978-3-030-96187-9 ISBN 978-3-030-96188-6 (eBook)
https://doi.org/10.1007/978-3-030-96188-6

This Springer imprint is published by the registered company Springer Nature Switzerland AG
The registered company address is: Gewerbestrasse 11, 6330 Cham, Switzerland

About the Book

Computational Social Science (CSS) is the science that investigates social and behavioral dynamics in both nature and society, through computer simulation, network analysis, and the science of complex systems. The Computational Social Science Society of the Americas (CSSSA) is a professional society that aims to advance the field of CSS in all its areas, from fundamental principles to real-world applications, by holding conferences and workshops, promoting standards of scientific excellence in research and teaching, and publishing novel research findings.

This volume is comprised of a selection of the latest research into CSS methods, uses, and results, as presented at the 2021 annual conference of the CSSSA. This conference was held virtually during November 4–7, 2021. What follows is a diverse representation of new approaches and research findings, using the tools of CSS and agent-based modeling (ABM) in exploring complex phenomena across many different domains. As such, readers will not only have the methods and results of these specific projects on which to build, but will also gain a greater appreciation for the broad scope of CSS, and have a wealth of case study examples that can serve as meaningful exemplars for new research projects and activities. This book, we hope, will appeal to any researchers and students working in the social sciences, broadly defined, who aim to better understand and apply the concepts of complex adaptive systems to their work.

Contents

About the Editors

Dr. Zining Yang is Senior Manager at Southern California Edison. She also works as Adjunct Professor at California State Polytechnic University, Pomona, and Associate Director at the TransResearch Consortium. She sits on the Board of the Computational Social Science Society of the Americas (CSSSA) and serves as Scientific Advisory Board Member for Human Factors and Simulations. She received her Ph.D. in Computational and Applied Mathematics and Political Economy from Claremont Graduate University. Her research interests include data analytics, machine learning, modeling and simulation, complex adaptive systems, agent-based models, and network analysis. She has published numerous times in the fields of computer science, economics, public policy, and political science. She has been identified as Outstanding Researcher by the government, worked on a National Science Foundation-sponsored project, and won multiple awards from various organizations, including the Ministry of Education of the People's Republic of China, International Social Computing, Behavioral Modeling and Prediction, and the International Institute of Informatics and Systemics.

Dr. Elizabeth von Briesen is Assistant Professor of Computer Science at Elon University and is Member of the board of the Computational Social Science Society of the Americas. She received her Ph.D. in Computing & Informatics from the University of North Carolina at Charlotte in 2020. Her research interests are focused on the study of complex adaptive systems using computational techniques, particularly with respect to social systems experiencing identity-based conflict. She primarily works with agent-based models and performs data mining and sentiment analysis to inform those simulations. Finally, in her current position, she strives to contribute toward an evolving undergraduate computer science experience through research, service, and high-quality, innovative teaching.

Effects of Assortativity on Consensus Formation with Heterogeneous Agents

Ece Mutlu and Ozlem Ozmen Garibay(✉)

Industrial Engineering and Management Systems,
University of Central Florida, Orlando, FL 32816, USA
{ece.mutlu,ozlem}@ucf.edu

Abstract. Understanding the consensus formation and exploring its dynamics play an imperative role in studies of multi-agent systems. Researchers are aware of the significant effects of network topology on the dynamical process of consensus formation; therefore, much more attention has been devoted to analyzing these dependencies on the network topology. For example, it is known that the degree correlation between nodes in a network (assortativity) is a moderator factor which may have serious effects on the dynamics, and ignoring its effects in information diffusion studies may produce misleading results. Despite the widespread use of Barabasi's scale-free networks and Erdos-Renyi networks of which degree correlation (assortativity) is neutral, numerous studies demonstrated that online social networks tend to show assortative mixing (positive degree correlation), while non-social networks show a disassortative mixing (negative degree correlation). First, we analyzed the variability in the assortativity coefficients of different groups of the same platform by using three different subreddits in Reddit. Our data analysis results showed that Reddit is disassortative, and assortativity coefficients of the aforementioned subreddits are computed as −0.0384, −0.0588 and −0.1107, respectively. Motivated by the variability in the results even in the same platform, we decided to investigate the sensitivity of dynamics of consensus formation to the assortativity of the network. We concluded that the system is more likely to reach a consensus when the network is disassortatively mixed or neutral; however, the likelihood of the consensus significantly decreases when the network is assortatively mixed. Surprisingly, the time elapsed until all nodes fix their opinions is slightly lower when the network is neutral compared to either assortative or disassortative networks. These results are more pronounced when the thresholds of agents are more heterogeneously distributed.

Keywords: Consensus · Degree correlation · Multi-agent system · Reddit

1 Introduction

Networks are representations of the connection patterns of complex systems in which entities might be proteins, individuals, economic goods, etc. Topological

Z. Yang and E. von Briesen (Eds.): CSSSA 2021, SPCOM, pp. 1–10, 2022.
https://doi.org/10.1007/978-3-030-96188-6_1

structures and different properties of diverse complex networks have been investigated for decades. Most importantly, the characterization of the mixing patterns of these network structures helps us to understand the evolutionary, functional and dynamic process of those complex systems [2]. If nodes in a network tend to associate with other similar nodes, this pattern is called assortative mixing (also known as *homophily*). The concept of assortativity is extensively studied since its introduction by Newman [18] in 2002. Although its application areas are diverse, the assortativity of a network is generally determined by the Pearson correlation coefficient between the degree distribution of its nodes. Previous studies show that non-social networks generally show a disassortatively-mixed pattern, i.e. metabolic pathways, protein-protein interactions, power-grid, World-Wide-Web [19] or yeast genes and proteins [15]; however, brain connections are assortatively mixed [12] despite its non-social property. Social networks, on the other hand, tend to be assortative, i.e. Facebook [22], Flickr, mySpace [11]; however, there are some exceptions such as disassortative mixing patterns on Twitter [24] and Youtube [11]. In such platforms where social networks are established, it is not surprising that people interact with other people similar to them. This similarity might be based on age, race, language, education or number of connections established. Fisher et al. argued that social networks are assortative only when they are built as a group-based network [8]. To either test the validity of this assumption and to understand the variability in the assortativity of social networks better, we evaluated the mixing patterns of three different subreddits and observed that assortativity values vary even in the different groups of the same platform. Although measuring assortativity will not give an idea about the variation of all entities in a network, it is useful in understanding the average mixing behavior of them; thus, it plays an important role in understanding the dynamics of epidemic spreading, signal connections, information diffusion or consensus formation in a system.

In this paper, we will be focusing on the effect of the average mixing behavior of heterogeneous agents on consensus formation. It is important to obtain a better understanding of consensus formation, since beliefs and opinions in social groups, including the society, constantly evolve as societal dynamics introduce paradigmatic shifts over time. Fashion trends, cultural changes, the rise and fall of political ideologies, marketing practices and technology innovations are good examples for these paradigm shifts leading to consensus formation. Simultaneous to these external influences in the society; internal communication patterns, such as online social networking activities of individuals, can also influence opinion formation, adoption, and dissemination of agents. For instance, it would be unexpected for an individual to adopt an opposing opinion in a network that predominantly supports another opinion. These internal communication patterns strongly depend on the network topology. Consensus formation of multi-agent systems agents have attracted researchers from many different disciplines. This concept is applied in many areas from spacecraft [23] to robotic teams [4]. There are applications of consensus strategies in decision-making, the polarization of people in the examples of political affiliation [13] and rumor spreading [5]. The multiplicity and diversity of uses necessitate a better understanding of the consensus formation process.

2 Assortativity in Reddit

As aforementioned, Twitter and Youtube show disassortative mixing patterns while Facebook, Flickr, and mySpace are assortative. This variability brings the following question that "Do the assortativity values of the different groups of the same social network vary?". The best example of the existence of different groups in a single platform is Reddit, in which authors can post their texts, URLs, images or videos, and/or add comments to other author's contents. Since contents are organized according to different subjects by user-created boards, which are called "subreddit" and thus, members can follow threads only if they are interested in it. Therefore, we collected all user activities in three different subreddits.

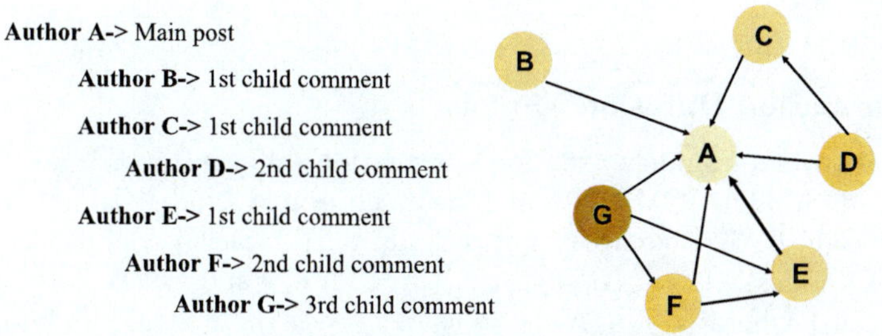

Fig. 1. A representation of Reddit network generation from a cascade.

Since there is no following or friendship in Reddit, we generated the Reddit network from the conversation patterns of users in subreddits. Suppose that we have a cascade in which one author posted (*Author A*) and 6 other authors added a comment (*Author B-G*) as shown in Fig. 1. Since commenter authors reply *Author A*'s post, we assume that *Author A* is followed by all other users in the network (same as Twitter, if you follow other users, you can see their tweets). Since *Author D* adds a comment to *Author C*'s comment but not *Author B*'s comment, we assume that *Author B* has no influence on *Author D*, and *Author D* follows *Author C* but does not follow *Author B*. In the last three comments of the cascade, on the other hand, *Author G* follows both *Author E* and *Author F* since cascade flows from parent to children comments continuously.

We generated three different networks from three cybersecurity-related subreddits from January 2015 to September 2017. Since we created our network based on an assumption, we first checked the power-law fitness of in-degree and out-degree distribution of the network. Many studies demonstrated that the majority of the social networks are scale-free networks [3]. High R^2 values[1] show that in-degree and out-degree of the generated Reddit network show a great power-law fitness, which supports the validity of our assumption. After generating networks, we calculated assortativity coefficients of these subreddits (Table 1). Motivated by the variability in the values even in the same platform, we decided to investigate the sensitivity of dynamics of consensus formation to the assortativity of the network.

Table 1. Details of Reddit datasets from 3 Subreddits

	Subreddit1	Subreddit2	Subreddit3
Number of user-generated contents	16,332,268	1,446,083	162,506
Number of unique users (*Author*)	751,561	119,088	33,762
Assortativity coefficient (r)	−0.0384	−0.0588	−0.1107

3 Method

3.1 Information Diffusion Models

There are two main models used for information diffusion, which are Independent Cascade Model (ICM) [9] and Linear Threshold Model (LTM) [10]. Suppose that $G\langle V, E\rangle$ is a complex network (graph), which is defined as the set of vertices (nodes) ($V = \{v_1, v_2, ..., v_n \mid n \in N*\}$) and edges between them ($V_{ij} = (v_i, v_j)$ where $(i, j \in N*; i \neq j, \forall i, \forall j < n)$). In ICM, individuals are assumed as bounded-rational, and their adoption can be predicted from the influence on each other. Here, each edge (V_{ij}) has its own influence probability (p_{ij}) which are previously known, and each node (i) has a single chance to activate its neighbor j with probability p_{ij}. As in the case of LTM, individuals adopt a new opinion only if a critical fraction of their neighbors have already adopted the new opinion; thus, every node (i) is associated with a threshold value (ϕ_i) to be activated in the next step.

Despite the necessity of assigning heterogeneity to thresholds in LTM, this heterogeneity is poorly-defined among researchers, which leads to extensive use of uniform [16,20,21] and binary [25] thresholds in many studies. Arguably, this assumption of homogeneous or binary thresholds is an oversimplification of reality and may produce misleading results. To remedy this oversimplification and thereby provide more holistic and possibly more accurate models, many studies employed more complex threshold functions such as the tent-like function [27], the truncated normal distribution function [14] or the sigmoid function [7]. In the current study, we employ LTM to understand the dynamics of opinion formation and control the threshold heterogeneity with a parameter N_{th} during the simulations.

[1] (0.9661, 0.9855), (0.9622, 0.9921) and (0.9559, 0.9778) for (in-degree, out-degree) of three subreddits, respectively.

3.2 Simulation Procedure

Although the main analysis here is to understand the effect of the assortativity of the network on the dynamics of consensus formation; we also investigated the effect of threshold heterogeneity to justify the robustness and scalability of the results. The heterogeneity of the agents is yielded by the diversity of their thresholds.

We first generated a random network using a configuration model, in which out-degree distribution is assigned as random numbers drawn from power-law distribution in the form of $\sqrt{N}x^{\gamma}$. Here, $\gamma = 3$ and N denotes the number of nodes in a network, that equals to 1000. Furthermore, we keep the in-degree as constant ($k_{in} = 17$), motivated by Dunbar number, i.e. individuals have a cognitive limit on the number of their social relationships [6]. This random network tends to be neutral (uncorrelated) or slightly disassortative. To tune the magnitude of the assortativity, we applied the Xulvi Brunet-Sokolov rewiring algorithm [26]. This algorithm chooses two linked node pairs at each time step, i.e. i, j and m, n where $A_{ij}, A_{mn} = 1$ and A denotes the adjacency matrix. Then, it orders these four nodes according to their degree, i.e. Suppose that $k_i < k_m < k_j < k_n$. To increase the assortativity, first two nodes and last two nodes i.e. $A_{ij}, A_{mn} = 0$ and $A_{im}, A_{jn} = 1$; to decrease the assortativity, first node with the last node and second node with the third node are rewired by destroying the previous linkage, i.e. $A_{ij}, A_{mn} = 0$ and $A_{in}, A_{mj} = 1$. This process continues until the desired assortativity is obtained. Note that, this algorithm does not change the overall degree distribution, thus mean a degree in the network; however, a rewired network may exhibit different geometrical and transport properties. The Xulvi Brunet-Sokolov algorithm considers only the out-degree distribution of the directed graph since the in-degree is constant.

In the next step, we initialized the opinions of individuals as a Bernoulli distributed random variable with an initial probability (p), i.e. the opinion of the node i (s_i) might equal to 1 with a probability p and equal to 0 with probability $(1-p)$. This probability value of p has a range of 0.2 to 0.8 in the current study.

The heterogeneity of the agents is yielded by the heterogeneity in their thresholds of adopting a new opinion. For this purpose, we assigned thresholds as a uniformly distributed random variable ($\phi_i \in unif(0.5, 1)$) throughout the interval defined by N_{th} which takes value from the set of $2, 5, 10, 100$ and increasing N_{th} yield more heterogeneity among agents. Thresholds of the agents are randomly assigned from the subset defined by:

$$\phi_i \in \{0.5, 0.5 + \frac{0.5}{N_{th} - 1}, ..., 0.5 + \frac{0.5(N_{th} - 2)}{N_{th} - 1}, 1\} \quad (1)$$

After generating the network and bringing its assortativity to the desired degree, initializing the opinions and assigning thresholds, we run the opinion change simulations. The process of updating opinions is as follows as in [17]:

1. Picking a node i randomly.
2. Calculating the weighted average of the opinions of its in-neighbors ($\bar{o}_i$). Here, weights are the multiple edges formed between node i and its neighbors.

3. Updating the opinion of node i (s_i) according to the criteria as follows:
   (a) **if** $s_i = 0$ and $\bar{o}_i - s_i > \phi_i$,
   then $s_i = 1$ in the next step.
   (b) **if** $s_i = 1$ and $\bar{o}_i - s_i < -\phi_i$,
   then $s_i = 0$ in the next step.

This Markovian chain is repeated until individuals fix their opinion.

4 Results

In the current study, we first aimed to analyze the effect of change in the network mixing pattern on the average opinion at steady state ($\bar{s}$). Therefore, after all the individuals fix their opinions in the network, we averaged their opinions by using the equation below:

$$\bar{s} = \frac{1}{N}\sum_{i}^{N} s_i(\infty) \tag{2}$$

where $s_i(\infty)$ is the opinion of node i at steady-state and N is the seed size, i.e. number of nodes in the network. We conducted our simulations to measure $\bar{s}$ as a function of the initial probability (p) with a varying assortativity (r) and varying threshold heterogeneity N_{th}.

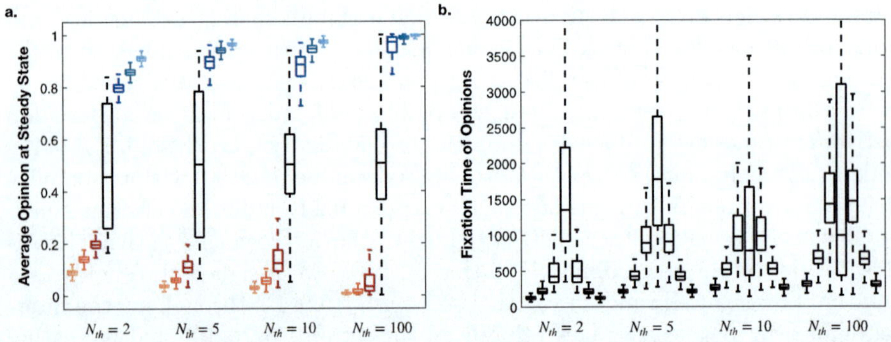

Fig. 2. a. Average opinion at steady state ($\bar{s}$) b. Fixation time of opinions (t_F) when $N = 1000$, $r = 0$ and $N_{th} = \{2, 5, 10, 100\}$. For each N_{th} seven boxplots show the variation in $p = 0.2, 0.3, ..., 0.8$.

Box plots in Fig. 2 show the distribution of repeated experiments on the average opinion at the steady-state ($\bar{s}$) (left) and the time elapsed until all individuals fix their opinion (t_F) (right) when $N = 1000$. We fixed assortativity coefficient of the network as neutral ($r = 0$) since many studies in the literature generates Barabasi's scale-free and/or Erdos-Renyi networks, which are tend to show uncorrelated mixing pattern. Each figure includes 4 groups of box plots, in which groups represent various N_{th}, and p varies from 0.2 to 0.8 in each group.

Since average opinion shows the sample mean of the opinions at steady-state, the values closer to either 0 or 1 show the dominance of one of the opinions, i.e. the system is more likely to reach a consensus. Figure 2a shows that p determines the dominance of the opinions; when $p \leq 0.4$, $\bar{s}$ take values close to 0, while $p \geq 0.6$ it approaches to 1. Furthermore, there is a clear asymmetry before and after $p = 0.5$ in all cases. As p closes to 0.5, system has mix of both opinions rather than reaching a consensus. Figure 2b, on the other hand, shows the relatively long duration of opinion change process when opinions are initialized halfway ($p = 0.5$). Although threshold heterogeneity of nodes in the system has a slight effect in the resulting average opinion when thresholds are out-degree dependent, we can conclude that the probability that the system reaches a consensus slightly increases as the threshold heterogeneity increases; however, the time elapsed until all nodes fix their opinions increases significantly with increasing N_{th}.

Figure 3, on the other hand, shows the effect of rewiring the network before opinion update simulations, when network is assortative ($r = 0.1$), neutral ($r = 0$) and disassortative ($r = -0.1$). All simulations are carried out at $N_{th} = 5$ and $N_{th} = 10$ for each mixing pattern to understand the moderator effect of the threshold heterogeneity. Results show that the increase in the assortativity coefficient of a network has a prominent effect on the average opinion at the steady state at both threshold heterogeneity. The system is more likely to reach a consensus when network is disassortatively mixed and this effect is more prominent when thresholds are more heterogeneous. Surprisingly, t_F reaches its minimum value when network is neutral. Figure 3b shows that bringing the network to the steady-state takes more time when degree distribution of its nodes either positively or negatively correlated. Again, this unexpected effect is more prominent when threshold heterogeneity is high.

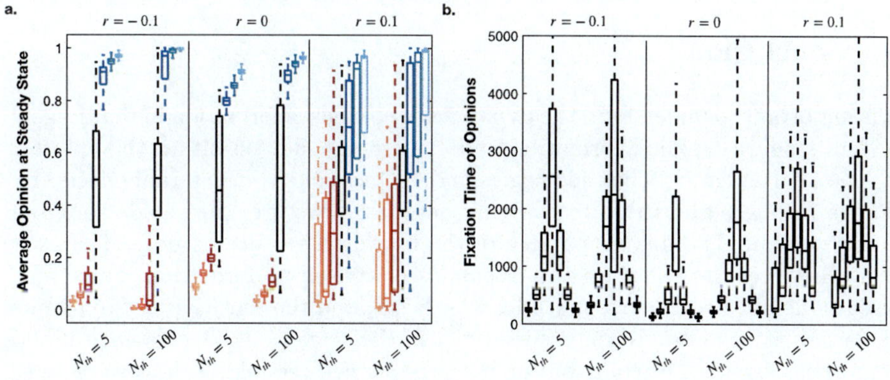

Fig. 3. a. Average opinion at steady state ($\bar{s}$) b. Fixation time of opinions (t_F) when $N = 1000$, $r = -0.1$ in dissortative, $r = 0$ in neutral and $r = 0.1$ in assortative mixing. For each $N_{th} = 5$ or $N_{th} = 100$, and seven boxplots show the variation in $p = 0.2, 0.3, ..., 0.8$.

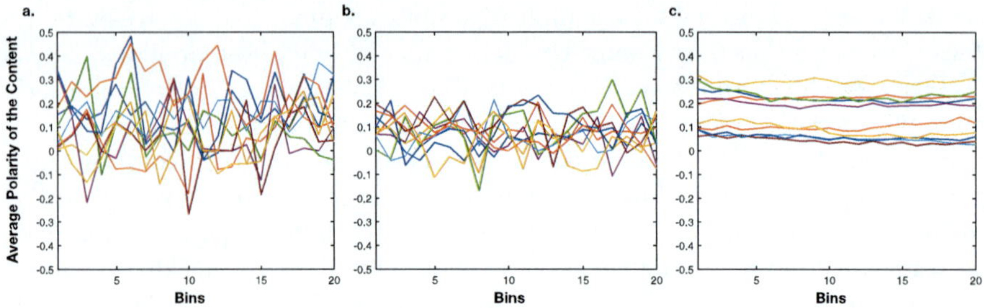

Fig. 4. Average polarization of the user-generated contents from 10 largest cascades in a. Subreddit1, b. Subreddit2 c. Subreddit3

To validate our results, we observed the opinion change of users in 10 largest cascade of three aforementioned subreddits. We considered the sentiment polarity, which is a float within a range $[-1.0, 1.0]$, for every user-generated content by using CLIPS's sentiment function [1] as a opinion towards the content of the discussion. Since each cascade have different size, we grouped (i.e. by averaging) the polarity scores into 20 bins after sorting them according to their time. We investigated the fluctuations in the polarity scores for each cascade in Subreddit1 ($r = -0.0384$), Subreddit2 ($r = -0.0588$) and Subreddit3 ($r = -0.1107$). Figure 4 shows the fluctuations in the average opinion at every time step in these subreddits. When network is less disassortative (close to neutral), opinion fluctuates excessively (Fig. 4a); while opinions show a relatively stable pattern when network is more disassortative (Fig. 4c). Since thresholds of users in Reddit are ambiguous but we concluded that threshold heterogeneity has a slight effect on reaching the consensus; we can argue that these real-world results match with our simulation results.

5 Conclusion

It is important to understand the dynamics of consensus formation in multi-agent system studies. Opinion formation and change depends on either the external effects, i.e. change in belief, ideologies and/or technology, or internal effects, i.e. change in interactions due to evolving network structure, thresholds to adopt a new opinion. The mixing pattern of the networks has been studied for many years, however, its effect on the dynamics of consensus formation is not analyzed in detail. In the current study, we investigated the sensitivity of dynamics of consensus formation to the assortativity of the network in the existence of heterogeneous agents. The contribution of the paper is three-fold: First, we created a Reddit network based on the conversation pattern of the users and validated our assumption with the power-law fitness of degree distributions. Our results show that Reddit topology can be created by using the post-comment relationships. Second, we explored the variability in the assortativity of sub-networks in the same platform. Although assortativity coefficients of many networks are available in the literature, none of the studies give information about Reddit network to our knowledge. For this purpose, we calculated the assortativity coefficients of the three subreddits and found that those values vary greatly, but all show disassortatively mixed behavior. Third, we examined the effect of assortative mixing in networks on the dynamics of consensus formation with multi agent-based simulations. During the simulations we tested the effect of assortativity coefficient of network, initial probabilities of the different opinions and threshold heterogeneity of the agents in the network structure. We concluded that the system is more likely to reach a consensus when the network is disassortatively mixed or neutral; however, the likelihood of the consensus significantly decreases when the network is assortatively mixed. Surprisingly, the time elapsed until all nodes fix their opinions is slightly lower when the network is neutral compared to either assortative or disassortative networks. Reaching the consensus is more likely but more time-consuming when thresholds of agents in the system are more heterogeneous. This slight effect of heterogeneity is observed every cases regardless of the mixing pattern of the nodes; however, its positive effect reaching a consensus is more pronounced when the system is disassortatively mixed. We also validated our findings with real world Reddit data.

Acknowledgment. This work is supported by grant FA8650-18-C-7823 from the Defense Advanced Research Projects Agency (DARPA). Also, we would like to thank Ivan Garibay for his valuable contributions that substantially improved this paper.

References

1. Pattern.en, December 2010
2. Allen-Perkins, A., Pastor, J.M., Estrada, E.: Two-walks degree assortativity in graphs and networks. Appl. Math. Comput. **311**, 262–271 (2017)
3. Aparicio, S., Villazón-Terrazas, J., Álvarez, G.: A model for scale-free networks: application to twitter. Entropy **17**(8), 5848–5867 (2015)
4. Arkin, R.C., Balch, T.: Cooperative multiagent robotic systems (1997)
5. Borge-Holthoefer, J., Moreno, Y.: Absence of influential spreaders in rumor dynamics. Phys. Rev. E **85**(2), 026116 (2012)
6. Dunbar, R.I.M.: Neocortex size as a constraint on group size in primates. J. Hum. Evol. **22**(6), 469–493 (1992)
7. Fink, C., Schmidt, A.C., Barash, V., Kelly, J., Cameron, C., Macy, M.: Investigating the observability of complex contagion in empirical social networks. In: Tenth International AAAI Conference on Web and Social Media (2016)
8. Fisher, D.N., Silk, M.J., Franks, D.W.: The perceived assortativity of social networks: methodological problems and solutions. In: Missaoui, R., Abdessalem, T., Latapy, M. (eds.) Trends in Social Network Analysis. LNSN, pp. 1–19. Springer, Cham (2017). https://doi.org/10.1007/978-3-319-53420-6_1
9. Goldenberg, J., Libai, B., Muller, E.: Talk of the network: a complex systems look at the underlying process of word-of-mouth. Mark. Lett. **12**(3), 211–223 (2001)
10. Granovetter, M., Soong, R.: Threshold models of diffusion and collective behavior. J. Math. Sociol. **9**(3), 165–179 (1983)

11. Hai-Bo, H., Wang, X.-F.: Disassortative mixing in online social networks. EPL (Europhys. Lett.) **86**(1), 18003 (2009)
12. Im, K., Paldino, M.J., Poduri, A., Sporns, O., Ellen Grant, P.: Altered white matter connectivity and network organization in polymicrogyria revealed by individual gyral topology-based analysis. Neuroimage **86**, 182–193 (2014)
13. Jin, C., Li, Y., Jin, X.: Political opinion formation: initial opinion distribution and individual heterogeneity of tolerance. Physica A **467**, 257–266 (2017)
14. Karampourniotis, P.D., Sreenivasan, S., Szymanski, B.K., Korniss, G.: The impact of heterogeneous thresholds on social contagion with multiple initiators. PloS One **10**(11), e0143020 (2015)
15. Lee, I., Kim, E., Marcotte, E.M.: Modes of interaction between individuals dominate the topologies of real world networks. PloS One **10**(3), e0121248 (2015)
16. Liu, Q.-H., Lü, F.-M., Zhang, Q., Tang, M., Zhou, T.: Impacts of opinion leaders on social contagions. Chaos: Interdisc. J. Nonlinear Sci. **28**(5), 053103 (2018)
17. Mutlu, E.C., Garibay, I.: The degree-dependent threshold model: towards a better understanding of opinion dynamics on online social networks. arXiv preprint arXiv:2003.11671 (2020)
18. Newman, M.E.J.: Assortative mixing in networks. Phys. Rev. Lett. **89**(20), 208701 (2002)
19. Newman, M.E.J.: Mixing patterns in networks. Phys. Rev. E **67**(2), 026126 (2003)
20. Singh, P., Sreenivasan, S., Szymanski, B.K., Korniss, G.: Threshold-limited spreading in social networks with multiple initiators. Sci. Rep. **3**, 2330 (2013)
21. Sprague, D.A., House, T.: Evidence for complex contagion models of social contagion from observational data. PloS One **12**(7), e0180802 (2017)
22. Ugander, J., Karrer, B., Backstrom, L., Marlow, C.: The anatomy of the Facebook social graph. arXiv preprint arXiv:1111.4503 (2011)
23. VanDyke, M.C., Hall, C.D.: Decentralized coordinated attitude control within a formation of spacecraft. J. Guidance Control Dyn. **29**(5), 1101–1109 (2006)
24. Wang, G., Wang, B., Wang, T., Nika, A., Zheng, H., Zhao, B.Y.: Whispers in the dark: analysis of an anonymous social network. In: Proceedings of the 2014 Conference on Internet Measurement Conference, pp. 137–150. ACM (2014)
25. Wang, W., Tang, M., Shu, P., Wang, Z.: Dynamics of social contagions with heterogeneous adoption thresholds: crossover phenomena in phase transition. New J. Phys. **18**(1), 013029 (2016)
26. Xulvi-Brunet, R., Sokolov, I.M.: Changing correlations in networks: assortativity and dissortativity. Acta Physica Polonica B **36**(5), 1431 (2005)
27. Zhu, X., Wang, W., Cai, S., Eugene Stanley, H.: Dynamics of social contagions with local trend imitation. Sci. Rep. **8**(1), 7335 (2018)

Quantifying Polish Anti-semitism in Twitter: A Robust Unsupervised Approach with Signal Processing

Peter A. Chew(✉)

Galisteo Consulting Group, Inc., 4004 Carlisle Blvd NE Suite H, Albuquerque, NM 87107-4566, USA
pachew@galisteoconsulting.com

Abstract. Anti-Semitism has long been a controversial topic in Poland, and remains so to the present day. Holocaust research continues in Poland today, attempting to face up to the truth about the nature and extent of any Polish complicity with the Nazis in the persecution and massacre of Jews during World War II. But such research faces significant headwinds in various forms, including popular Polish nationalism as well as the passage in 2018 by the current Polish nationalist government of a law against defaming the 'Polish nation' in connection with the Holocaust. A recent court case against a Poland-based Holocaust researcher again brought this to the fore, lighting up the Polish Twittersphere with chatter on the topic. In this paper, we set out to analyze this chatter, to shed light – not on any past Polish anti-Semitism – but on *present-day* Polish attitudes. Along the way, we argue that unsupervised techniques from signal processing are the ideal tool for wading into such a politically and emotionally charged topic. Such techniques simply reflect, in condensed and quantifiable form, what is in the data. It is then up to the human analyst to draw the appropriate conclusions.

Keywords: Twitter · Poland · Anti-semitism · Unsupervised analysis

1 Background

In this paper, we demonstrate a novel way of thinking about computational social analysis in a controversial area: anti-Semitism in present-day Poland. The algorithms are not new; what is novel is their placement within an analytic framework, and we will argue that this placement ideally positions the analyst seeking to make sense of data which by its nature is contentious, and best leverages what state-of-the-art analytic techniques offer. But first, to help contextualize why unsupervised text analytics is so appropriate for the problem we set out to solve in this paper, we provide a detailed social background on Poland.

Poland, today a relatively small nation-state of 38 million on the eastern edge of the European Union, has occupied an outsize role in the history of the last century, owing to its position directly between the giants of Europe, Germany and Russia. Indeed, several centuries ago, the Polish-Lithuanian Commonwealth in fact occupied a much larger

Z. Yang and E. von Briesen (Eds.): CSSSA 2021, SPCOM, pp. 11–22, 2022.
https://doi.org/10.1007/978-3-030-96188-6_2

area, encompassing most of present-day Poland, Ukraine, Belarus, Lithuania, Latvia and Estonia, a fact of which Poles today, eager to emphasize their 'great-nation' status, will remind visitors. Poland is a liminal area of Europe: ethnically Slav, like its eastern neighbor Russia, but overwhelmingly Catholic, unlike its Orthodox neighbors to the east; one of the most Westernized parts of the Soviet bloc during the Cold War, but now one of the most socially conservative (and in the eyes of some, illiberal) European Union states.

The area of the Polish-Lithuanian Commonwealth has also historically been the home of a large concentration of Ashkenazi Jews, possibly as many as 700,000 [1]. This number is almost certainly significantly lower today; many Jews either fled or were killed during the World War II Holocaust, the epicenter of which was in Poland. While the Holocaust resulted first and foremost from Nazi ideology, it may be fair to say that the Holocaust also exposed a Polish 'raw nerve': attitudes towards Jews in Poland, where the Catholic Church often preached a doctrine of 'supersessionism' [1], have historically also been far from simple. This is something which the Polish Pope Jan Paweł II implicitly acknowledged by kneeling and praying in 1978 at Auschwitz [2]. But despite Jan Paweł's attempts to mend fences, the issues of old keep resurfacing, for example in a 2017 tweet by Jacek Międlar, a Catholic priest based in southwestern Poland, in which Międlar associates 'healthy Polish nationalism' with 'hatred of the Talmud' (Fig. 1).

Fig. 1. An example of modern Polish/Catholic anti-Semitism

[1] Supersessionism is the doctrine that the Christian Church replaced the Jews as God's chosen people from the time of Jesus Christ, and that since then, being Jewish has no relevance to salvation.

The complexity of Polish-Jewish relations is perhaps nowhere better illustrated than in the history of the Holocaust. The most notorious concentration camps, such as Auschwitz, where the Nazis attempted to exterminate Jews, are located in Poland and anyone who wants to see the chilling evidence can do so, as this author has. Yad Vashem, the Holocaust remembrance center, counts by far the most Poles of any single nation among the 'Righteous Among the Gentiles' [3], because many Poles were instrumental in saving Jews from the Nazis, even at personal risk. But both sublime bravery, and the basest betrayal, are possible within the same nation and even by a single individual; and a focus of modern Holocaust research in Poland has been to establish the facts without fear or favor. As Barbara Engelking, co-author of one such study [4], has said, 'there are no black-and-white situations in Holocaust history… history is more complicated than [the ruling nationalist party in Poland] would like it to be' [5].

That this is even today a sensitive subject in Poland is clear. For example, in her work [4], Engelking claims that Edward Malinowski, mayor of a Polish town in World War II, 'was complicit in the death of several dozen Jews who were hiding in the forest and were turned over to the Germans'. (Engelking has also acknowledged [5] that Malinowski also saved the life of one Jewish woman, Estera Drogicka.) For her assertion about complicity, Engelking was sued in a Warsaw court in 2019 by Malinowski's niece, who in turn was supported by 'Reduta Dobrego Imienia', a legal defense fund set up to defend the name of the Polish nation. All this came in the context of a 2018 law passed by the nationalist Polish ruling party, Law and Justice (Prawo i Sprawiedliwość), outlawing 'blaming the Polish government or people for complicity in the Nazi Holocaust'[2]. In February 2021, the court rejected any claim against Engelking for damages, on the basis that this could have a 'chilling effect on research', but ordered Engelking to apologize for 'inaccuracies' in her book. While acknowledging she did get some factual details wrong, Engelking has asserted that these were not really those that the plaintiff or court focused on, and the case is expected to be appealed [5]. Meanwhile, though, the court's verdict elicited a significant number of strong reactions in the Twittersphere, mostly from within Poland, raising the question that we will attempt to answer here: what do Poles really think about all this? What can Twitter data tell us about Polish nationalism and even anti-Semitism today?

2 Prior Literature

We are not aware of any research using social media analytics into the specific question of Polish attitudes towards the Holocaust. However, there is of course a plethora of research analyzing Twitter data. Studies attempting to answer the research question we pose would typically approach it as a sentiment analysis problem (e.g. [6, 7]). Twitter data would be collected and analyzed, perhaps for 'positive' versus 'negative' sentiment, determined on the basis of tweets (or words from the dictionary) labeled by human annotators. Essentially, human judgment is made far 'upstream' in the analytical process, before going through a sentiment analysis 'black box', and sentiment judgments are rendered by a machine 'downstream'. Judgment is inherently subjective, not to mention

[2] For text of law (in Polish) see https://sip.lex.pl/akty-prawne/dzu-dziennik-ustaw/zmiana-ustawy-o-instytucie-pamieci-narodowej-komisji-scigania-18685855/art-1.

the non-trivial problems of (1) dealing with words whose sentiment varies significantly according to context, and (2) establishing what precisely positive or negative sentiment is directed towards. This raises the question of how human judgment and bias plays into computational algorithms. Could automation possibly unintentionally magnify effects of bias, or lead in other ways to unintended consequences? If such sentiment analysis presents its results as 'quantitative science', but is in reality based on upstream subjective human decisions that might not generalize, might this even raise ethical questions about the way in which results are presented and relied upon?

In this connection, it is worth considering what CIA veteran Richards Heuer wrote about bias; the foreword to his book 'The Psychology of Intelligence Analysis' reminds us that 'the pitfalls the human mental process sets for analysts cannot be eliminated; they are part of us. What can be done is to train people how to look for and recognize these mental obstacles, and how to develop procedures designed to offset them'. In an arena as sensitive as Polish attitudes towards Holocaust studies and anti-Semitism today, and with the obvious legal pitfalls that have already been highlighted, one cannot afford to allow computational algorithms to lead an analyst to conclusions that are insufficiently qualified with appropriate caveats, in the way that good intelligence analysis should be. In high-stakes intelligence analysis, lives may depend upon an accurate reading of intercepted communications; here, lives may not be at stake, but careers, reputations, monetary damages and jail time may be, as the Polish legal landscape has made clear. As Heuer acknowledges, bias is inevitable, but we must at least have ways to be transparent about where the bias is. It is not clear how this can happen effectively where a sentiment analysis 'black box' builds these biases in from the very start in ways which are hard to quantify.

The contribution of our work starts by returning to basics: we focus squarely on answering Heuer's (and, we may infer, the intelligence community's) technology-agnostic call to 'develop procedures designed to offset [people's mental obstacles]'. Our goal is to re-imagine how this process can work from start (the social media data) to finish (useful, potentially actionable quantitative analysis), finding the best available analytical tools such that processes are repeatable, any bias is eliminated from the analytical process, or at least introduced as far downstream in the process as possible. Bias should also be introduced in a way where it is not hidden in a 'black box', and responsibility for bias is clearly devolved to, and controlled by, the human analyst whose conclusions can be challenged in the normal way that would happen in the intelligence community. As Heuer says, bias is unavoidable – it can even help lead to analytical breakthroughs – but it must be made transparent.

Much of our own previous work has focused on applications of unsupervised algorithms which have found many productive uses in everyday life (search engines, recommender systems), and it turns out that such algorithms can fit the bill very well in helping an analyst make sense of controversial social media data. In such a charged area as Polish anti-Semitism, bias might easily lead an analyst to see what he or she wants to see, but our goal should be to see things as they are, no matter if the truth is ugly. Action and decision-making are surely more effective when informed by truth than rose-colored conclusions. We show how unsupervised techniques used in signal processing can help get us to that point.

3 Unsupervised Summarization to Fulfill Heuer's Vision

Singular Value Decomposition (SVD) underlies Latent Semantic Analysis (LSA) [9], the use of which for topic modeling and summarization on unstructured text has been recognized for a long time [10, 11]. The use of LSA for text summarization, in fact, is a special case of SVD's use as a signal processing algorithm. The essence of signal processing generally, and SVD in particular, is to take a noisy input and automatically separate out the input's 'principal components' of signal. The input is not changed in this process, but rather *factorized*, such that the components can be recombined faithfully to reconstruct the original input. One way of looking at SVD is that it finds groups of 'similar' things in the input, presenting them in the output such that those groups that account for the largest signal patterns are ordered first. In matrix computation terms, and represented graphically, the SVD factorization is as follows:

$$X = U \cdot S \cdot V^H = \sum_{i=1}^{r} s_i \cdot u_i \cdot v_i^H$$

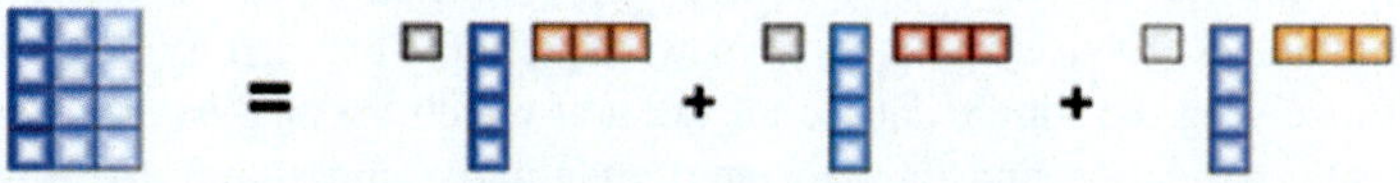

Fig. 2. Graphical representation of SVD

Each triad of vectors in the factorization thus represents a 'component of signal', with those ordered first that (in a statistical sense) are most important in the source data (Fig. 2).

The application to text summarization is as follows. In a corpus of documents consisting of sentences (or other chunk) of text, one can imagine that there might be a number of sentences similar to one another, although perhaps differing in a few words here or there. Examples might be the two headlines in Table 1.

Table 1. Example 'similar' sentences

1	Holocaust Scholars Engelking and Grabowski Ordered to Apologize in Libel Case
2	Poland orders Holocaust scholars to apologize in case that could muzzle research

One way of thinking about summarization is as eliminating redundancy, but at the same time highlighting a selection of representative-example sentences that most describe the full panoply or dimensions of what is in the source text, but also ensuring that the summarization accurately reflects that some groups of sentences have more members than others. (For example, there might be more sentences about the 'order to apologize' than about the 'chilling effect on research').

LSA offers a way to rethink this problem in signal processing terms, simply letting the statistics (rather than any human intuition) drive the result. In LSA, each triad of vectors represents a component of signal orthogonal to every other. Crucially, within each triad, one of the vectors can 'score' each chunk of text with respect to that component of signal. To obtain the summarization, therefore, we can simply list the top most 'representative' chunk of text for each of the r triads. The value of r can be varied in what is known as a 'truncated SVD', such that a smaller r provides a more condensed summarization (but less complete with respect to all the semantic dimensions of the input). Or a larger r might be selected by an analyst with more time, who wants to delve into more detail. Either way, it is guaranteed (as a feature of SVD) that those 'components of signal' that are most widespread in the source text will be ordered first, and that the factorization is the best possible approximation to the original corpus, in a statistical (least-squares) sense, given the selected value of r. Further, an artifact of LSA is that we can numerically quantify the similarity of any chunk of text to any other (by computing a cosine between the chunks' vectors v). This is one method we can use to estimate how large each group is – how much of the original corpus it accounts for.

From all this it should be apparent that we are already a very long way towards fulfilling Heuer's call to develop 'procedures to offset [bias]'. LSA summarization offers a way, in effect, for an analyst to view a large corpus of noisy data in 'miniature'. But the summarization is driven entirely mathematically, simply by what is in the data. From the point of receiving the source data to the point at which the analyst can examine the results (the summarization and its associated quantitative measures), there is little to no room for bias being unwittingly introduced through heuristics built into any kind of 'sentiment analysis' or other algorithms. The analyst can rest securely in the knowledge that the process, if repeated, will yield the same results, and represents the best possible approximation to the original data in a mathematically-explainable sense. He or she is then free to apply his or her own extrinsic background knowledge, bias, insights, etc., or not. What happens from this point on is simply what regularly occurs (and can be challenged) within the intelligence community, but the point is that it is part of ways our approach can be *used*, rather than part of the algorithmic 'black box'.

4 The 'Polish Holocaust Research' Dataset

4.1 Data Collection

Almost invariably, popular sentiment surrounding issues such as the Holocaust or anti-Semitism does not exist in a vacuum. It arises, to a greater or lesser extent, as a reaction to events in real-time. This is why, for example, the Anti-Defamation League has reported correlations between the timing of Israeli-Arab violence in the Middle East and anti-Semitic attacks in Western countries [12]. The same is true in the Twittersphere. It is not that hateful attacks or tweets do not reflect hate that pre-existed in hearts and minds before the events transpired; rather, we believe that the events simply serve as a catalyst to reveal attitudes people already had.

When the outcome of the controversial court case involving Barbara Engelking was announced in Poland on February 9, 2021, and the case subsequently hit the international news headlines, therefore, it seemed like a good time to take the temperature of the Polish

Twittersphere. On February 22, 2021, we used the Twitter Developer API to cast a specific but wide net: we ran a query to return all available tweets mentioning ‘Engelking’. This returned a respectable 5,796 tweets. Note that the search term ‘Engelking’ is a specific enough name that we could be confident that substantially all the returned results would relate to Barbara Engelking and the recently-concluded court case; but also, since ‘Engelking’ is a proper name, the query did not limit the results to tweets originating in Poland, or in Polish. Nevertheless, since the topic was of particular interest in Poland, unsurprisingly, an overwhelming majority of the tweets returned *were* in Polish (Fig. 3). Additionally, the search query ‘Engelking’ was carefully chosen so as to avoid (to the extent possible) introducing bias ‘upstream’, that is, before application of any computational algorithms. By including only the (rather distinctive) name, we hoped to avoid skewing the returned results towards those that might be positive or negative.

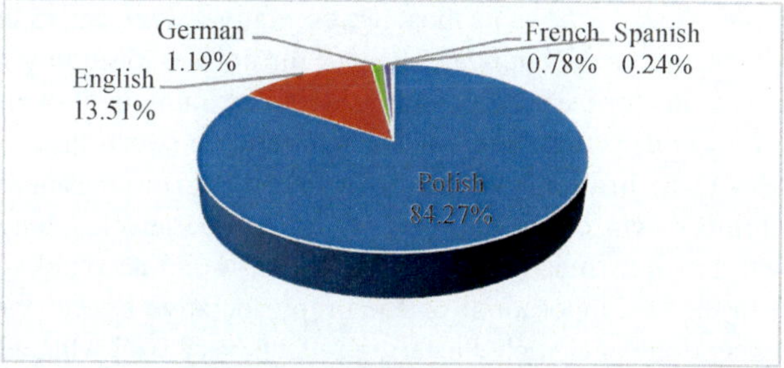

Fig. 3. Distribution of ‘Engelking’ Twitter data by language

As a result, even in just the data collection, by posing a Twitter API query in an open-ended but specific way, we already learn something valuable: in which languages the topic is most of interest. Specifically, we confirm, even before applying any algorithmic analysis, that the Engelking case did seem to touch a nerve among the Polish-speaking community. Not all topics, by any means, garner as much attention as the almost 6,000 tweets (within the space of only around 72 h) evidenced.

4.2 Data Processing

To process the data and gain a summary-level view, we construct a term-by-tweet matrix, a standard step in information retrieval, which, it seems, could hardly be questioned in terms of the possibility of introducing bias. A tweet is a unitary whole corresponding to its author’s intended communication. The terms are simply the words the author chose. Tweets are also short enough that they force the author to stick, generally, to a single thought within the tweet. This is a built-in boon for analysis, in that it generally reduces the possibility of tweets belonging to multiple semantic clusters simultaneously, making the results easier to interpret.

The matrix is then weighted according to pointwise mutual information (pmi) between term i and tweet j. If $p(i|j)$ represents the probability of term i within tweet *j*, and p(*i*) the overall probability of term *i* in the entire corpus, etc., then:

$$\mathrm{pmi} = \log\left(\frac{\mathrm{p(i|j)}}{\mathrm{p(i)}}\right)$$

This measure is also solidly grounded in information theory and is driven only by the observed statistics within the collected data, not by any heuristics, intuition or subjectivity. It should not be surprising that good theory leads to empirically better results on similar problems in information retrieval, as has been demonstrated with respect to pmi [13]. Here again, as in every step, we attempt to stay true to Heuer's vision by adopting scientifically justifiable and bias-neutral techniques.

We then factorize the matrix using a truncated SVD to obtain the top 100 'topics', as described in Sect. 3 above. For each triad of output vectors, we find the most representative tweet and the 20 most representative terms and tabulate these results. It is possible for two topics to have the same most representative tweet, and in those cases, we condense the results to eliminate duplicates. (Since the point is to summarize the corpus, duplication adds no value for the analyst.) We also find the number of tweets which have a similarity of 0.8 or greater to the top most representative tweet, based on the cosine similarity measure (again, justifiable because the SVD factorization is in an orthonormal space). 0.8 is an arbitrary cut-off and another value could be selected, but the point here is just to get a sense of how much of the corpus each 'topic' accounts for: how many other tweets are in the 'neighborhood' of each representative tweet? We have found, with almost no exceptions, that such clustering reflects very well what a human would intuitively consider 'similar' texts, especially when based on PMI combined with SVD as described above, since the point of PMI is to measure how much mutual information documents and terms share. An example of texts that would be clustered in this way are the following (note the common 'distinctive' terms 'Holocaust', 'scholars', 'Engelking', 'Grabowski', 'apologize', etc.):

Holocaust Scholars Engelking and Grabowski Ordered to Apologize in Libel Case https://t.co/A8eW0BI3fn
And what a chilling/muzzle effect on political, historical science, culture (literature, cinema) will it have? Holocaust Scholars Engelking and Grabowski Ordered to Apologize in Libel Case https://t.co/nz7niSkFa2
Poland orders Holocaust scholars to apologize in case that could muzzle research Court rules Barbara Engelking and Jan Grabowski disseminated false information and harmed the honor of a man they said helped kill Jews in WWII; his niece brought the suit https://t.co/KipHz2jrHw

The output of all the above is such that it can be pasted into an Excel spreadsheet. The *final* step, performed by the human analyst, is to annotate these results manually based on critical thinking, and sorting into categories, specific to the problem at hand. What is the language of each topic (as already mentioned, not all will be Polish)? This is a relatively trivial analytical step and could well be automated, though can quickly be accomplished manually on condensed data. What is a summary of each representative tweet in English? (Again, this can be semi-automated using, for example, Google Translate.) Is the tweet positive or negative towards Engelking and her co-authors? In the analyst's judgment, does the tweet express anti-Semitic sentiment? An overview of the final output after human analysis looks as shown in Fig. 4 (not intended to be legible in detail, but included to show impressionistically the major contours of the output), where the blue rectangle indicates the human's annotation. Color-coding in the first column is by language of tweet (and by extension, cluster).

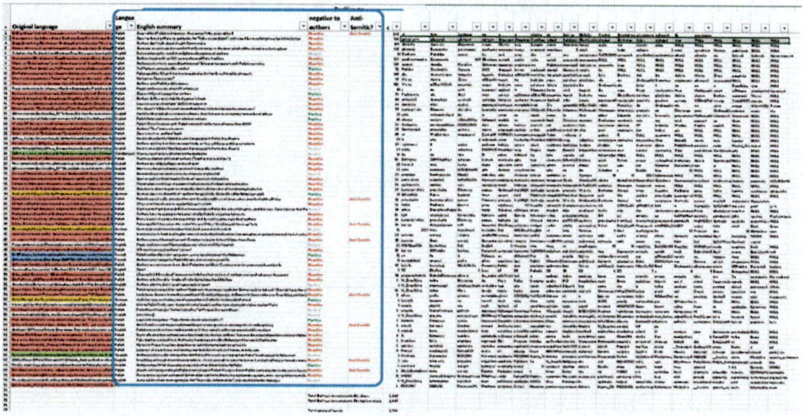

Fig. 4. Overview of analytical output. Annotation by human analyst is circled in blue.

To summarize, we have taken close to 6,000 tweets in their original languages, summarized them in a bias-neutral fashion so that analyst can see the 'forest' instead of the 'trees'. Now, the analyst has to read just 71 tweets to get a picture of the most significant thoughts expressed in the whole corpus. In fact, we know from the 'neutral', automated part of the analysis that over 50% of the tweets in the corpus have a similarity of at least 0.8 to at least one of these 71 tweets. Sentiment analysis takes place on the *back end*, fully under the control of the analyst who can tailor the analysis to the problem at hand, but it is an extremely manageable problem for a human, with 71 rather than 6,000 tweets to read. It is still possible, however, to extrapolate in a scientifically justifiable way from the analysis of the condensed corpus back to the full corpus, knowing how many tweets have a high level of similarity to the analyzed 'summary' tweets. With this in mind, what kind of results can be obtained?

5 Results

We have already presented in Fig. 3 the distribution of tweets by language, showing that over 5 in 6 tweets retrieved using the 'Engelking' query are in Polish. Using the human back-end analysis (of which we shall show some examples below), we can obtain a sense of how many tweets were positive or negative with respect to Engelking and her co-authors, stratified by language of tweet (Fig. 5):

A greater level of negativity towards Engelking et al. within the tweets in Polish (compared to other languages) tends to support the idea that Engelking's book was taken by nationalist Poles as an affront to their narrative about the Polish character.

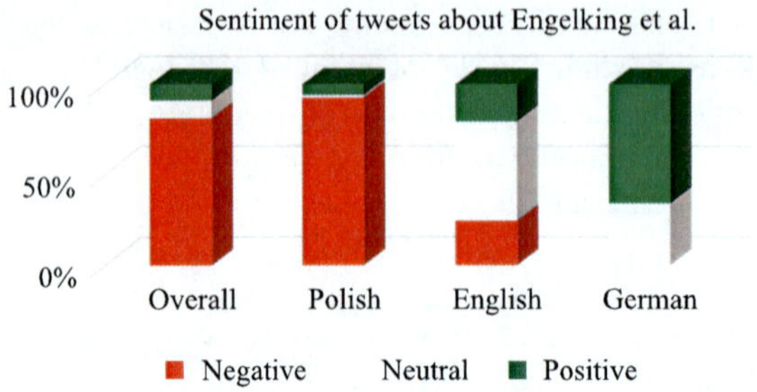

Fig. 5. Proportion of tweets, by language, which are negative towards Engelking et al.

A similar analysis can be performed with respect to anti-Semitic sentiment stratified by language (Fig. 6):

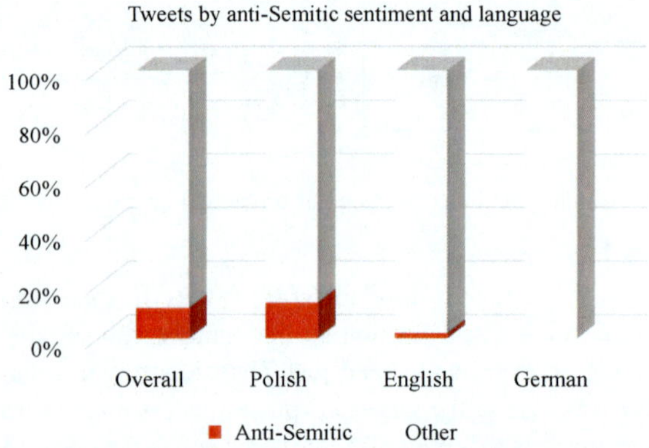

Fig. 6. Proportion of tweets, by language, which express anti-semitism

This analysis suggests that anti-Semitic content is by no means in the majority of tweets, but at over 15% of tweets in Polish (over 84% of the corpus), it is worryingly high. Anti-Semitism also runs considerably higher in Polish-language tweets than in other languages (1.8% in English and 0% in German).

So how did we (as the analysts) determine if a tweet was anti-Semitic, positive or negative? (Recall, these judgments are made by a human, not by any algorithm.) We used a fairly straightforward criterion for anti-Semitism: if the text of the tweet made any negative generalization about Jews, or referenced such historical generalizations, then we considered it anti-Semitic. Examples from the data: 'Jews robbed Polish aristocracy, they weren't the ones robbed'; 'Poles wouldn't have seen the end of the war if they'd had to wait for Jews' help'; a sideswipe at Gazeta Wyborcza (a newspaper) calling it 'Gazeta Żydowska' ('Jewish Newspaper') for its pro-Engelking position; 'Engelking's accusing Poles is to make money'; 'Jewish entities operating in Poland carrying out Polish-financed anti-Polish campaign for years'. Note that some of the anti-Semitism here is very subtle, though instantly recognizable to the informed. One would need to know about historical anti-Semitic narratives (such as 'Jewish money') or contextual information (such as Gazeta Wyborcza's pro-Engelking position) to recognize them. It is doubtful that any automated sentiment analysis could do this, illustrating why human post-analysis, after 'bias-free' signal processing, is so desirable. Negativity (or positivity) towards Engelking can be manifested just as subtly. Negativity, of course, is not necessarily anti-Semitism; one can be critical of a position without resorting to generalizations. Examples of tweets negative towards Engelking: 'Brave Leszyńska proved manipulation and deceptive propaganda of authors' (Leszyńska is the niece who brought the lawsuit, again requiring contextual knowledge); 'Authors have to apologize to Leszyńska for falsely accusing her uncle' (the tweet assumes the accusation is false). Neutral tweets are those that stick to the facts (which in this case are complex), for example: 'Press conference on Filomena Leszczyńska's civil libel lawsuit'; 'Putin/Kremlin praise "historical policy" of Prawo i Sprawiedliwość' (neutral because the tweet, through use of quotation marks, does not take a position); 'Judge said authors must apologize for "imprecise information", rejected claims for damages'. Example of tweets positive towards Engelking (also requiring contextual knowledge): 'Lawsuit is dangerous - "false history leads to bad policy"'; 'Polish nationalist right-wing govt. wants to reinterpret the Holocaust'.

6 Conclusion

In this paper we have demonstrated a way of completely rethinking sentiment analysis within an automated text-analytic system. Our method still achieves the efficiencies of automated text analysis, but places responsibility for complex judgment calls back in the hands of the analyst, where we believe this responsibility belongs. The key is that we avoid building in biases and assumptions far 'upstream' in an analytic system and can therefore remain much truer to CIA veteran Richards Heuer's vision of 'procedures to offset [mental obstacles and pitfalls of bias]'. The role of automation here is simply to reflect what is in the data, in condensed form, making it easy for an analyst to process and understand the 'forest', using his or her full faculties of critical thinking and reasoning and tailoring them carefully to the problem at hand.

This way of thinking about sentiment analysis is, we believe, particularly necessary when dealing with high-stakes judgment. In Poland today, saying the wrong thing to challenge the nationalist orthodoxy can land researchers with fines or jail time. In some intelligence applications, people can die as a result of flawed analysis. It is therefore crucially important to know exactly where biases might be introduced into analysis. As Heuer says, bias is inevitable and it can even be made to work to advantage in the analytic process. But reasoning and conclusions must be transparent enough to be able to be challenged. Our approach places complex reasoning back in the hands of humans where it belongs, while still using automation as a force-multiplier for the human.

Applying this methodology to Twitter data, and with relevant data, we are able to get a reasonably clear picture of sentiment in Poland today. This picture is, to be sure, subject to some caveats. Twitter users are not necessarily representative of the entire population, either in Poland or other countries. Twitter data can also be manipulated, for example by bots. But regardless of what may have been true in the past, *today*, it seems there is still a significant undercurrent of anti-Semitism among Polish speakers – in as much as

15% of content. It seems some of the demons of Poland's history and past are still to be dealt with. For this to happen, facing up to the truth – however ugly or controversial it might be – is essential. And unsupervised text analytics techniques, in line with Heuer's vision, offer a way to help us do that.

References

1. Pogonowski, C.: Poland: a Historical Atlas. Hippocrene Books, New York (1989)
2. Jewish Virtual Library. Pope John Paul II: Relations with Jews and Israel (2005). https://www.jewishvirtuallibrary.org/pope-john-paul-ii-relations-with-jews-and-israel. Accessed 4 June 2021
3. Yad Vashem: Names and Numbers of Righteous Among the Nations - per Country & Ethnic Origin (2020). https://www.yadvashem.org/righteous/statistics.html. Accessed 4 June 2021
4. Engelking, B., Grabowski, J. (eds.): Dalej jest noc: Losy Żydów w wybranych powiatach okupowanej Polski. Polish Center for Holocaust Research, Warsaw, Poland (2018)
5. Szymczak, J.: Po wyroku na badaczy Zagłady. Prof. Engelking: "Nie czuję się winna" (2021). Oko.press https://oko.press/po-wyroku-na-badaczy-zaglady-prof-engelking-nie-czuje-sie-winna/. Accessed 4 June 2021
6. Uyheng, J., Carley, K.: Characterizing network dynamics of online hate communities around the COVID-19 pandemic. Appl. Netw. Sci. **6**, 20 (2021)
7. Pang, B., Lee, L., Vaithyanathan, S.: Thumbs up? Sentiment classification using machine learning techniques. In: Proceedings of the Conference on Empirical Methods in Natural Language Processing (EMNLP), Philadelphia, July 2002, pp. 79–86 (2002)
8. Heuer, R.: The Psychology of Intelligence Analysis. Center for the Study of Intelligence, Central Intelligence Agency, Langley, Virginia (1999)
9. Landauer, T., Foltz, P., Laham, D.: Introduction to latent semantic analysis. Discourse Process. **25**, 259–284 (1998)
10. Papadimitriou, C., Raghavan, P., Tamaki, H., Vempala, S.: Latent semantic indexing: a probabilistic analysis (postscript). In: Proceedings of ACM PODS, pp. 159–168 (1998). https://doi.org/10.1145/275487.275505. ISBN 978-0897919968
11. Srivastava, A., Sahami, M.: Text Mining: Classification, Clustering and Applications, p. 108. CRC Press, Boca Raton (2009)
12. Anti-Defamation League. Antisemitic Incidents Reported to ADL Increase Sharply During Israel-Hamas Conflict (2021). https://www.adl.org/blog/antisemitic-incidents-reported-to-adl-increase-sharply-during-israel-hamas-conflict. Accessed 7 June 2021
13. Chew, P., Bader, B., Helmreich, S., Abdelali, A., Verzi, S.: An information-theoretic, vector-space-model approach to cross-language information retrieval. Nat. Lang. Eng. **17**(1), 37–70 (2011)

On Modeling Evolution in Continuous Spaces

Robin Clark and Steven O. Kimbrough(✉)

University of Pennsylvania, Philadelphia, USA
kimbrough@wharton.upenn.edu

Abstract. Mendelian genetics, with its model of particulate, discrete genes, has been a power tool of analysis, as well as an attractive one that has drawn many researchers in cultural evolution. Many forms of cultural evolution, however, involve continuous variation over a space that is not easily quantized. In this paper, we look at continuous variation directly, not through the lens of particulate encodings, but rather from the perspective of mechanisms that generate continuous behavior (or phenotype). We study the degree to which we can come to understand cultural evolution in continuous space directly. We propose a spatial or landscape representation framework for evolution in continuous space of clustered behavior. The paper then presents four models of how clustered behavior patterns may evolve on their landscapes, including drift, selection, separation, and merging. These may be seen as building blocks for a comprehensive program of modeling cultural evolution.

Keywords: Cultural evolution · Mechanisms of evolution · Agent-based modeling · Evolution of vowels

1 Introduction

Mendelian genetics, with its model of particulate, discrete genes has been a power tool of analysis, as well as an attractive one that has drawn many researchers in cultural evolution. Notice, though, that many forms of cultural evolution involve continuous variation over a space that is not easily quantized. Furthermore, we have seen little compelling evidence that the quanta of cultural evolution have a physical, or at least testable, realization. In this paper, we propose to look at continuous variation directly, not through the lens of particulate encodings, and study the degree to which we can come to understand cultural evolution in continuous space directly.

Human height at adulthood, for example, varies continuously and it is possible to identify subpopulations that systematically differ in average height. The apparent continuous variation can in principle be explained by a combination of

CSS 2021 https://computationalsocialscience.org/conferences/css2021/. File: css2021-continuous-evolution.pdf.

Z. Yang and E. von Briesen (Eds.): CSSSA 2021, SPCOM, pp. 23–42, 2022.
https://doi.org/10.1007/978-3-030-96188-6_3

the trait's being controlled by multiple genes that in the aggregate are expressed in a noisy fashion. Different populations would be posited to have different ensembles of height-determining genes. Alternatively, height could be regulated by the interaction of genes and the environment, epigenetically. Of course, realized heights may be caused by a mixture of the two causal systems and there is no reason to assume that the two kinds of mechanisms should be in constant proportions across different traits. It will be useful to name the two fundamental approaches. We call the first the *collection of particles* or *particulate* form of explanation and the second the *machine* or *mechanism* form of explanation. We note that either mechanism can be realized by organismal development based on fundamentally particulate genetic material. We also note that human height, like many other continuously varying traits whether in biological or cultural systems, displays discernible clustering. Some populations, for example, are on average taller than others, and vowels may occupy different areas of sound spectrum in a common language. Taking the latter as an example, two languages may share a phonetic category, for example the vowel [i] as in English "meet." However, speakers of one language might produce the vowel differently than speakers of the other language, say by having a higher second formant. This is exactly the case of Italian and Yoruba. See Ladefoged (2006); Levinson and Dediu (2010).

Our evolutionary modeling investigations are focused mainly on social and cultural evolution. Evolution of vowel pronunciation is our model system for the sake of this short paper. Vowel evolution, besides being complex, is an excellent model for cultural evolution studies. It manifests continuous expression and evolution, there are multiple vowel systems in existence in the world's languages, displaying different numbers of vowels, and in every vowel system the constituent vowels occupy distinct regions in a continuous and connected space. These and other considerations strongly suggest a large role for a mechanism approach to modeling evolution of vowel systems. They are, moreover, representative of many cases of cultural evolution. The challenge then is to find a suitable kind of approach for modeling and explaining these kinds of systems. Our purpose in this paper is to present and develop such an approach.

Space limitations hardly permit a full development of the topic. Instead, our aim is to address the complexity and cultural evolution communities by (1) presenting the possibility of a non-particulate form of modeling and evolution for complex systems and social evolution, and (2) communicating from the ground up how this may be done, with the aims (i) of presenting easily replicable and extendable ideas, (ii) of demonstrating how simple building blocks (basic procedures) can be assembled to model explanatory mechanisms for social and cultural evolution, and (iii) of stimulating discussion of possibilities.

The paper is organized as follows. In Sect. 2, "Conceptual Model of Continuous and Clustered Cultural Evolution," we state more precisely our conceptual approach or framework for thinking about evolution on continuous landscapes (or spaces). Section 3, "Background on Vowels," provides essential background for the application area that has most directly motivated our concept development: the evolution of how vowels are pronounced. Evolution of this sort has been recorded by linguists and is accepted as a scientific fact. Because theoretical explanation of it remains unsettled, the topic calls for modeling, which has barely begun. Section 4, "Model 1: Single Vowel, Continuously Varying," models both random drift and directional selection for a single vowel in a population of communicating individuals. Section 5, "Model 2: Simple Selection," illustrates directional selection resulting from the presence of an environmental attractor that probabilistically censors random utterances the more they are distant from the attractor. Section 6, "Model 3: Preference for Separation," models how closely realized vowels (or concepts) may evolve to be more distinctly separated. Section 7, "Model 4: Avoidance and Merger," models a two-vowel system buffeted by two forces: one as in model 3 promotes separation of vowels (concepts), the other promotes their merger. What emerges from individual choices is compromise of the two forces in play. Section 8, "Model 5: PROBE AND ADJUST," models a learning machine that operates in continuous space. Finally, Sect. 9, "Discussion and Conclusion," tersely summarizes the paper, outlines the significance of the approach and findings, and points towards future research.

2 Conceptual Model of Continuous and Clustered Cultural Evolution

Figure 1, "Schematic landscape for social/cultural disposition," is useful for communicating our conceptual approach to, and for thinking about, social/cultural (hereafter cultural) evolution. The figure presents a static picture of a 2-D landscape, and should be seen as merely a special case of what may be an arbitrarily dimensioned space. At any given time, a point on the landscape (space) represents the possible behavior of an individual or group of individuals, as described by the dimensions x and y. The colored shapes represent clusterings of behavior. The figure displays three distinct clusterings, indicated by the colors red, green, and blue.

To construct a concrete example, x and y might be responses to survey questions, e.g., on attitudes towards gay marriage, x, and degree of alarm about climate change, y. (We are assuming the responses are measured on a continuous scale, but this is hardly necessary.) The data when plotted indicate three groups, or clusters, of respondents. Note that the boundaries of the clusters would be imposed. For example, they could represent regions capturing 95% of the data when an ellipse is superimposed on the data. Individual data points could well be spread throughout the landscape and there is no requirement that the clusters not overlap.

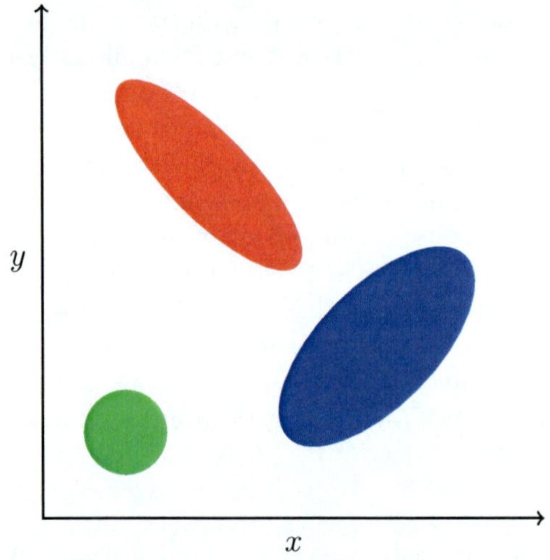

Fig. 1. Schematic landscape for social/cultural disposition.

Points arising:

1. This conceptualization, Fig. 1, is extremely general, if not universal, for traits or properties (x and y in the figure) of interest for cultural evolution.
2. Under this conceptualization, cultural evolution occurs when the disposition of clusters on the landscape changes. The study of cultural evolution is then about documenting and explaining such changes, ideally providing a causal account.
3. Biological evolution is often described on a particulate model: evolution is constituted by changes in gene frequencies.
4. It is at best a rough approximation to explain continuously varying behavior (as in Fig. 1) with a purely particulate model. Even in biological evolution, continuous behavior (phenotype) is largely explained by varying production from mechanisms that control the expression of genes. Often, much of what is evolved are mechanisms, genetically based of course, for responsive controlling of transcription and translation of so-called coding genes. Biologists understand very well the centrality of discovering the mechanisms of genetic control for explaining phenotypes and their evolution.
5. On analogy with biological evolution, cultural evolution is often (but not universally; see (Mesoudi 2011) for a review) conceived of as changes in frequencies of particles. The term *meme* is popularly applied to such hypothetical entities (Dawkins 1976; Blackmore 1999). This approach has not been as productive as might be hoped. As many have noted, scientific identification of memes is problematic, and explanations in terms of memes are sparse. The concept is now in large part relegated to humor and popular culture (Clark 2019).
6. The utility of memes as an explanatory principle is further diminished, or at least circumscribed, by the fact that, as in Fig. 1, the behavior to be studied exhibits continuous variation.

These considerations strongly suggest the importance, in the study of cultural evolution, of exploring responsive mechanisms that generate cultural behavior. They and the reasons behind them are the impetus for this paper. It will be important in any such research program to identify simple mechanisms of general applicability that can explain macro-behavior of cultural dispositions as in Fig. 1. Before reporting our findings in this regard, we turn in the next section to describing the context cultural evolution that most directly motivated for us this line of research, the evolution of vowel pronunciation.

3 Background on Vowels

The acoustic properties of vowels are the result of the resonating characteristics of the vocal tract; shaping the vocal tract using the articulators changes the timbre of the sound (Ladefoged 2001; Patel 2008). Figure 3 shows the formant structure for three vowels: /i/ is in "heap," /u/ is "hoop" and /ɑ/ as in "hop." The x-axis of the graph is time and the y-axis is frequency in Hertz. The formants are dark bands clustered at certain frequencies; it is this distribution of energy that gives the vowels their characteristics (Fig. 2).

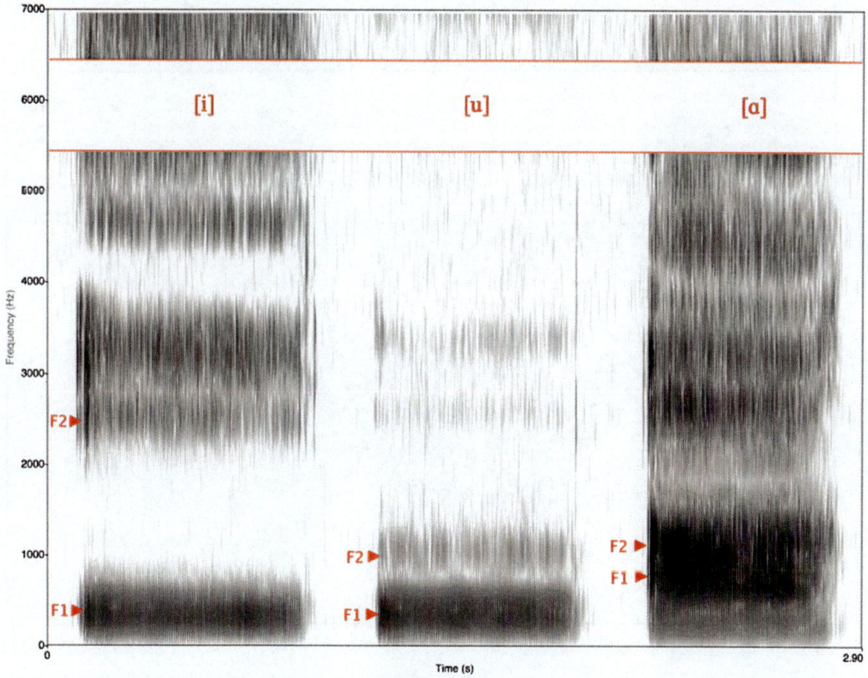

Fig. 2. Spectrogram show three vowels in English: /i/, /u/, and /ɑ/, Formants F1 and F2 are indicated in red. (from Wikipedia)

Figure 3 shows the distribution of a collection of vowels as produced in a free narrative by a native speaker of American English. The vowels are charted by the frequency of a normalized F1 on the x-axis and a normalized F2 on the y-axis. Although individual vowel utterances cluster together–the vowel [æ] as in "bat" is clustered in the upper left-hand corner–but they overlap with other categories–æoverlaps with the [ɛ] found in "bet" and the [ə] found in "but" (refer to Fig. 3).

Speakers identify vowels based not just on their F1 and F2 characteristics, but also on the aspects of their power spectrum, as well as top-down information provided by the possible lexical items being spoken. This top-down information can be inferred from the syntax and semantics of the utterance. Nevertheless, we can see that the space in which vowels occur is continuous, that the vowels form coherent categories and that these categories can overlap significantly. We will take vowel systems to be the outcomes of evolution over continuous space. They are particularly useful examples since data about vowels is copious, can be measured instrumentally, and represented numerically. All these features make them ideal for studying time series of evolution. See Labov (1994) for work in sociolinguistics that traces the time series of evolving vowels, although the perspective is rather different than that taken here.

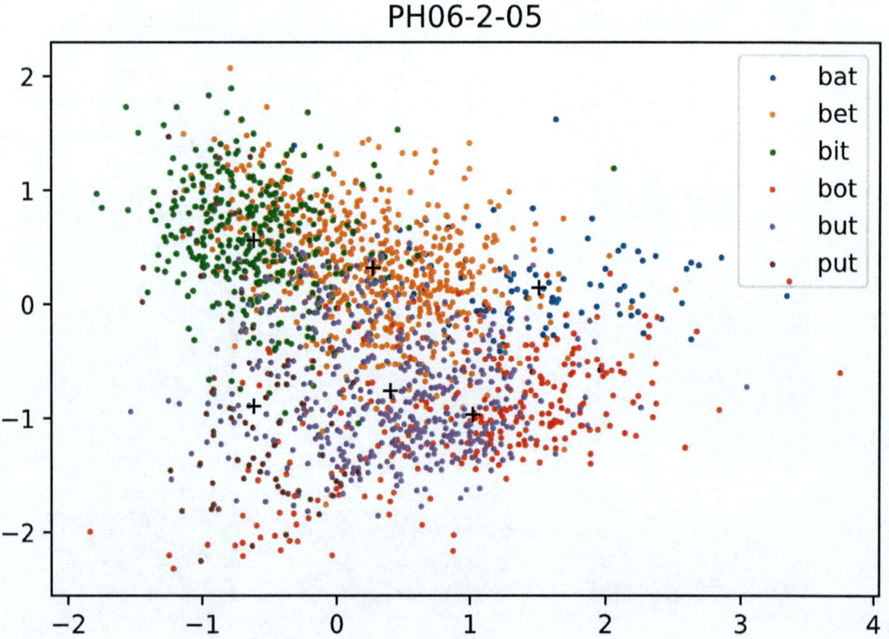

Fig. 3. Lax vowels charted by Normalized F1 and F2 for one speaker of Philadelphia English

4 Model 1: Single Vowel, Continuously Varying

Our first, most basic model is implemented in EvolutionInContinuousUnivariate.nlogo (available from https://github.com/stevenokimbrough/sokpapers/tree/master/2021/ContinuousEvolution-CSS2021 as well as the other models described in the paper). A run is initialized with `num-turtles` agents, each of which is endowed with an exemplar list of `num-exemplars` exemplars (of the modeled vowel or other entity). These in turn are independent draws from a normal distribution with mean 0 and standard deviation 1.5 (in the default case on which we focus; but both values may be varied at will).

The running of the model is a simple process. There are two modes of running: under drift and under influencers. Under drift, at each tick of the clock, an agent is picked at random to "speak". The agent calculates the mean and standard deviation of the items in its exemplar list (suggesting memory of recent utterances by other agents). The agent then draws a new value from a normal distribution with the mean and standard deviation of its current exemplar list. This new value is inserted into the end of the exemplar list of a random agent and the first item in that agent's list is deleted (forgotten). This completes the action until the next tick of the clock.

Under influencers, at each tick of the clock a parameter value `influencer-rate` is checked. If `influencer-rate mod ticks = 0` then an unnamed influencer makes a random draw from the initializing normal distribution (in the default case) This new value is inserted into the end of the exemplar list of a random agent and the first item in that agent's list is deleted (forgotten). This completes the action until the next tick of the clock. Otherwise, the program reverts to the drift procedure.

We track two principal statistics: `meanmean`, the agent population mean of the individual agent exemplar means, and `meansd` the agent population mean of the individual agent exemplar standard deviations. We note that the quantitative study of the microevolution of vowel systems is still in its early days. Despite this, we have some evidence regarding the properties of such change, thanks largely to the work of William Labov, who has built a large corpus of Philadelphia speech (Fruehwald 2013; Labov et al. 2013). The evidence indicates that, indeed, there is drift, although its character needs to be studied carefully. Furthermore, some variants seem to undergo selection, their movement being uncharacteristic of simple drift. We show below that the continuous model can also handle well-known processes of language change like splits, where phonologically conditioned variants of one vowel move apart and become two vowels, and mergers, where two vowels move together and become one vowel.[1] Finally, we can show the effect of language leaders ("influencers") on vowel change; this kind of effect is ubiquitous in cultural evolution. We can, then, using simple means account for a variety of

[1] For example, in northern cities in the United States some vowels have shifted and merged so that "bus" and "boss" are pronounced identically. Equally, in some dialects of 19th century British English, the vowels in "isle" and "oil" came to be pronounced identically, although they later split again and returned to their pronunciations.

evolutionary changes that affect both vowel systems and, more generally, cultural artifacts.

4.1 Under Drift Alone

We find that the behavior of the population falls into distinct patterns, or phases (Vorobeychik et al. 2014), each of which is a form of random drift because there are no directional forcing forces at work. We present representative configurations that reliably fall into the indicated phases.

1. (4 agents, 15 exemplars each) Drift until loss of variance: The mean of the exemplar means drifts arbitrarily above or below the initial value of 0 and the mean of the exemplar standard deviations goes to 0, indicating a complete loss of variation in the population. This is characteristic of small numbers of agents, 4 or fewer, and small to medium-sized (up to about 50) exemplar lists. Figure 4 plots a representative run.

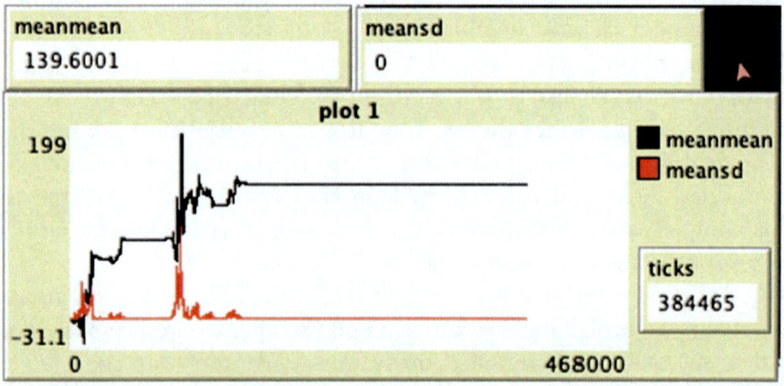

Fig. 4. Representative run: 4 agents, 15 exemplars each.

2. (50 agents, 50 exemplars each) Drift and growth of variance: The mean of the exemplar means drifts fairly rapidly above or below the initial value of 0, while the mean of the exemplar standard deviations grows inexorably. In a typical run, with 50 agents, each with 50 exemplars, the mean of the exemplar means drifts to below -5000 in less than 1,200,000 ticks. Figure 5 shows a plot from a typical run.

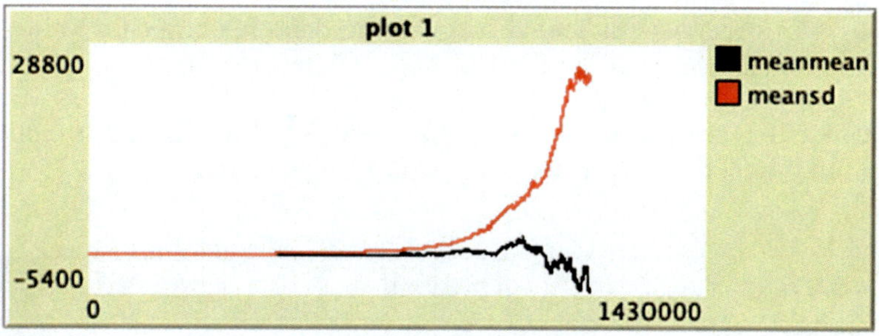

Fig. 5. Drift and growth of variance illustrated. meanmean = mean of the exemplar means. meansd = mean of the exemplar standard deviations. A representative run: (50 agents, 50 exemplars each).

3. (200 agents, 50 exemplars each) Subdued drift and relentless growth of variance: The mean of the exemplar means drifts slowly above or below the initial value of 0, while the mean of the exemplar standard deviations grows inexorably. This typically happens with large numbers of agents (e.g., 200) and medium or larger numbers of exemplars (50 or more). See Fig. 6 for the plot of a representative run. Notice that the larger population (200 agents) slows the rate of drift, as opposed to the rate in Fig. 5 with 50 agents.

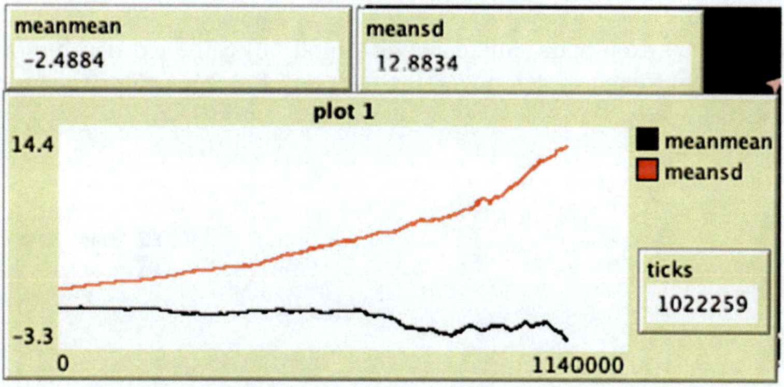

Fig. 6. A representative run: (200 agents, 50 exemplars each).

4.2 Under Influencers

Clark and Kimbrough (2015) modeled the emergence of linguistic patterns or phases in a population resulting from the influence of so-called "language leaders" (Labov 1994). Their model was one of evolution in a continuous space. Here we investigate whether a more general and thoroughly continuous model can also exhibit pattern formation caused by influencers, e.g., language or social leaders. Unless otherwise indicated, we use the default parameter values from the previous subsection.

1. (4 agents, 15 exemplars, influence with mod 50) Punctuated stability. See Fig. 7 for the plot of a representative run. Compare with Fig. 4.

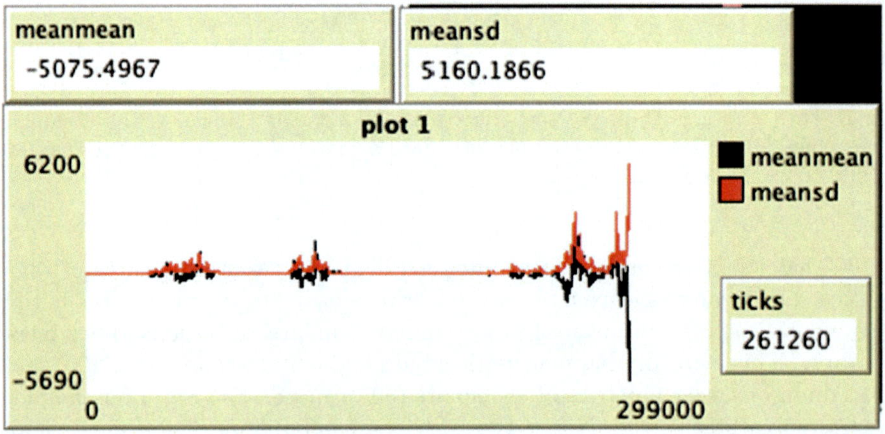

Fig. 7. A representative run with 4 agents, 15 exemplars each, influencer utterance every 50 ticks. Halted when the absolute value `meanmean` exceeded 5,000.

2. (50 agents, 50 exemplars, influence with mod 50) Subdued drifting of both meanmean and meansd. See Fig. 8 for the plot of a representative run. Compare with Fig. 5.

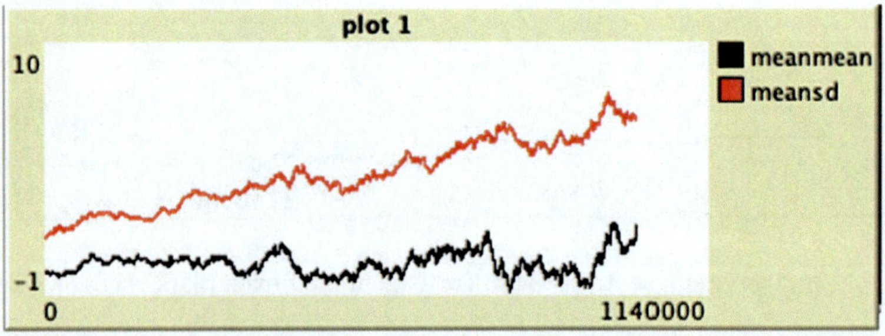

Fig. 8. 50 agents, 50 exemplars each, influencer utterance every 50 ticks: 1000000 tick run.

3. (200 agents, 50 exemplars each, influence with mod 50) Constrained drifting of meanmean near 0, slow increase in meansd. See Fig. 9 for the plot of a representative run. Compare with Fig. 6.

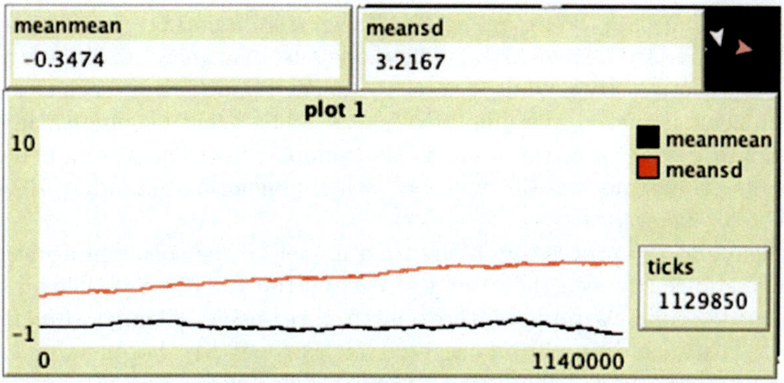

Fig. 9. Representative run: 200 agents, 50 exemplars each, influencer utterance every 50 ticks.

4. (200 agents, 50 exemplars each, influence with mod 50 and mean = 10) The single change from the previous experiment is that here when influencers draw an utterance they draw with a mean of 10 instead of 0. What we find is gradual approach to the influencer mean followed by drift about that mean, and increase in meansd followed by apparent drift. See Fig. 10 for the plot of a representative run.

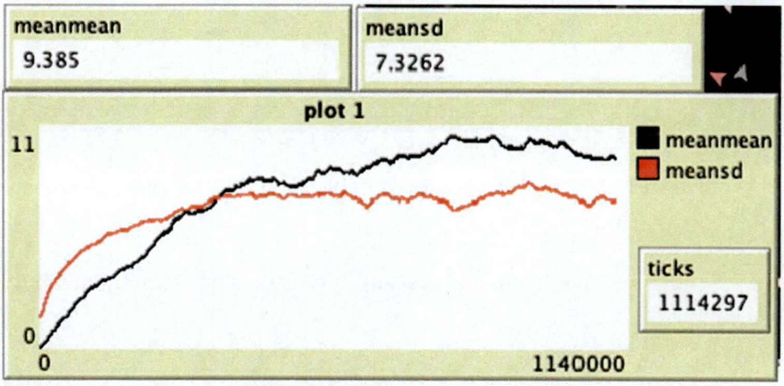

Fig. 10. Representative run: 200 agents, 50 exemplars each, influencer utterance every 50 ticks, with mean = 10.

5 Model 2: Simple Selection

Model2SimpleSelection.nlogo, available from the authors, illustrates directional selection. As in model 1, each agent is initialized with `num-exemplars` exemplars, and in the default case the exemplars are drawn from a normal distribution with mean 0 and standard deviation 1.5. The model also has an `IdealPoint` variable that drives selection. By default it is set to 10. As in model 1, at each tick of the clock, an agent is picked at random to "speak". The agent calculates the mean and standard deviation of the items in its exemplar list. The agent then draws a new value from a normal distribution with the mean and standard deviation of its current exemplar list.

At this point the processing differs from model 1. The speaking agent calculates the distance between the value it drew and the `IdealPoint`. This distance is used to determine a probability of acceptance, using a logistic function. A draw with distance of 0 will be accepted with probability 1; otherwise, as the distance increases the probability of acceptance decreases. Figure 11 shows the relevant part of the logistic acceptance function. A draw with a distance of 27 has a probability of acceptance of 0.497. With the probability determined via the logistic function, this new value is inserted into the end of the exemplar list of a random agent and the first item in that agent's list is deleted (forgotten). This completes the action until the next tick of the clock.

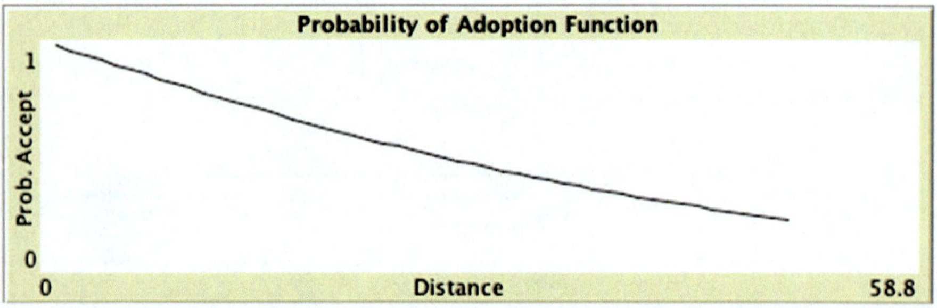

Fig. 11. Model 2 logistic acceptance function (in part).

Figure 12 plots results from a representative run with 100 agents having 20 exemplars each.

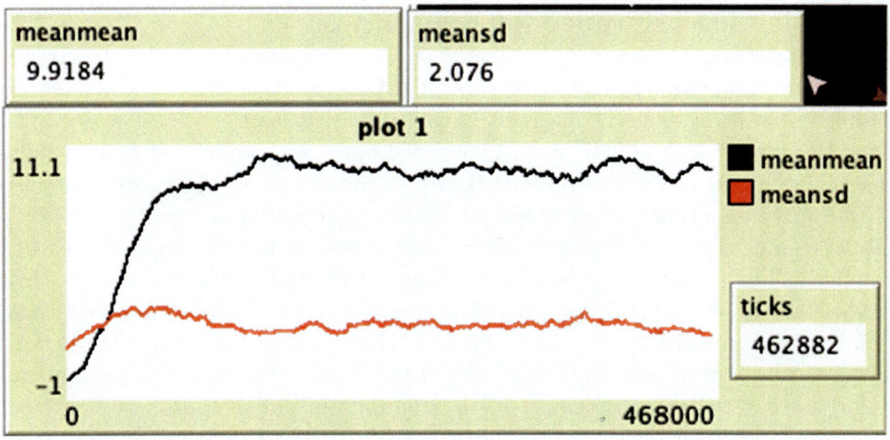

Fig. 12. Simple selection with 100 agents and 20 exemplars each.

In evidence is a pattern: the `meanmean` value increases steadily but noisily towards the `IdealPoint` value of 10, then fluctuates slightly at a value just below it; and the `meansd` value increases slowly until the `meanmean` approaches the `IdealPoint`, after which it declines very slowly and noisily. With 200 agents having 50 exemplars each, the pattern repeats itself with much less noise. See Fig. 13.

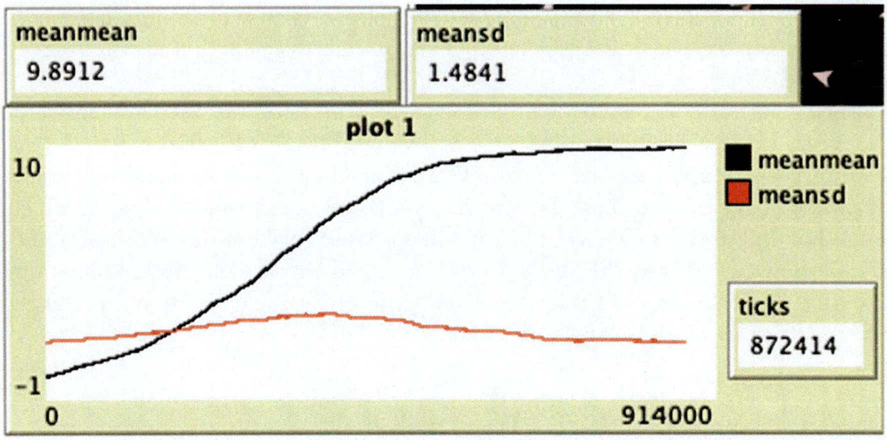

Fig. 13. Simple selection with 200 agents and 50 exemplars each.

6 Model 3: Preference for Separation

Model3SimpleSelection2Values.nlogo (available from the authors) illustrates gradual separation of two "vowels". Remarkably this emerges from local decisions. Hearers are told which vowel is uttered and they probabilistically decide whether to add it to their exemplars, based on proximity to their exemplars.

Initialization occurs as in models 1 and 2, except that now there are two vowels, v and v1. Inevitably, `meanmeanv` (the mean of the exemplar means for vowel v) will by chance differ from `meanmeanv1` (the mean of the exemplar means for vowel v1), even though all of the exemplars in the entire population are drawn from the same normal distribution. Also, inevitably some agents by chance will initially be organized contrary to the overall population. For example, `meanmeanv` could be greater than `meanmeanv1` in the population, but `exemplars-mean` (the mean of the exemplars for vowel v) could be less than `exemplars1-mean` (the mean of the exemplars for vowel v1) for any particular individual agent.

At each tick of the clock, a random agent is chosen to be speaker. The agent randomly chooses which vowel to utter, estimates the mean and standard deviation from its own exemplars for the vowel in question, and then draws an utterance from the corresponding normal distribution. The hearer accepts the utterance depending on a probabilistic calculation. The utterance's vowel (v or v1) is available to the hearer. Realistically this may be inferred from context. The closer the utterance is to the mean of the hearer's exemplars for that vowel, compared to the mean of the hearer's exemplars for the other vowel, the more likely the hearer will accept the utterance as a new exemplar. If the hearer accepts the utterance for the vowel, then as in the other models the utterance is added to the end of the exemplar list for the vowel in question and the first (oldest) exemplar in that list is deleted. This completes the action for the current tick period.

Figure 14 plots the results of a representative run. At initialization `meanmeanv` (the mean of the exemplar means for vowel v) was -0.114 while `meanmeanv1` (the mean of the exemplar means for vowel v1) was 0.016. After about 30,000 ticks (utterances elicited) the exemplars for the two vowels have separated, after which point they fluctuate somewhat but remain separated with a difference of about 4.75. Notice as well that the standard deviations of the two exemplars remain in lock step. Both are reduced from an initial expected value of 1.5 to a fluctuating but steady value of about 0.25.

7 Model 4: Avoidance and Merger

Model4ConflictSelection2Values.nlogo (available from the authors) illustrates interplay of merger and separation of two "vowels" in response to conflicting forces, separation as in model 3 and avoidance, introduced here.

Hearers are told which vowel is uttered and they probabilistically decide whether to add it to their exemplars, based on proximity to their exemplars (as in model 3), with a bias to reject utterances that increases with proximity to

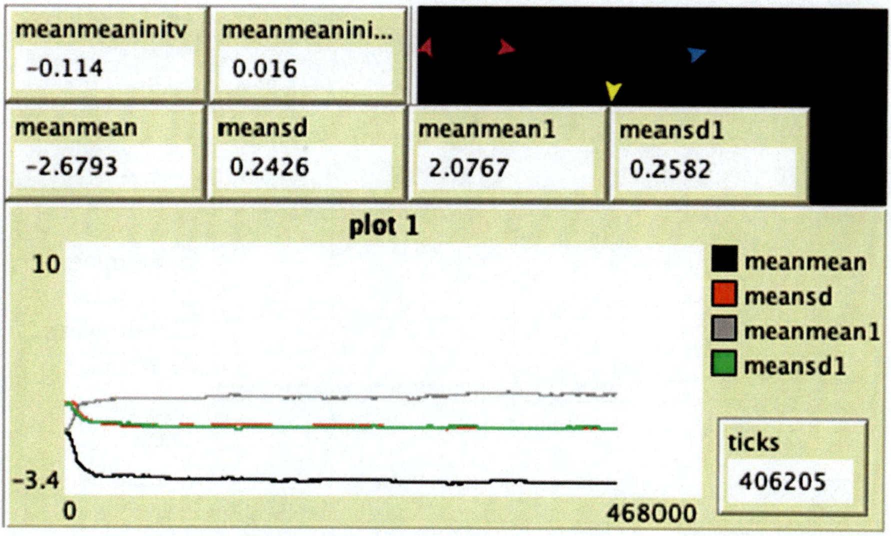

Fig. 14. Model 3: Simple section for separation of 2 concepts: 37 agents and 20 exemplars each.

either of two avoidance points. By default the avoidance points are set at −10 and 10. In a representative run, Fig. 15, exemplars for vowel 1 are drawn from a normal distribution with mean −4.0, while exemplars for vowel 2 are drawn with a mean of 4.0. We see in the figure (and in general in other runs) that under the parameter and decision settings used the two populations of exemplars begin far apart (differences of means of about 8) then asymptotically move towards each other for a fluctuating difference of about 2.6. Thus, the `meanmean1-meanmean2` difference is greatly reduced it is not eliminated. The two forces interact to produce a degree of merger, albeit an incomplete one.

Remarkably once more the behavior emerges from local decisions.

8 Model 5: bProbe and Adjust

Models 1–4 have served to demonstrate that a number of basic change (or evolutionary) behaviors on a continuous landscape can be explained by very simple mechanisms and without recourse to particulate combinations and mixings. Whether anything like these mechanisms obtains in a particular case is of course completely an empirical matter. Our purposes in this paper are propaedeutic: we aim to establish the substantial possibility of explaining cultural evolutionary phenomena with a non-particulate basis, relying instead on mechanisms and or machines that generate behavior. To this end, our final model in this paper—Model 5, Probe and Adjust—introduces a mechanism for agent learning of continuous behavior in continuous spaces.

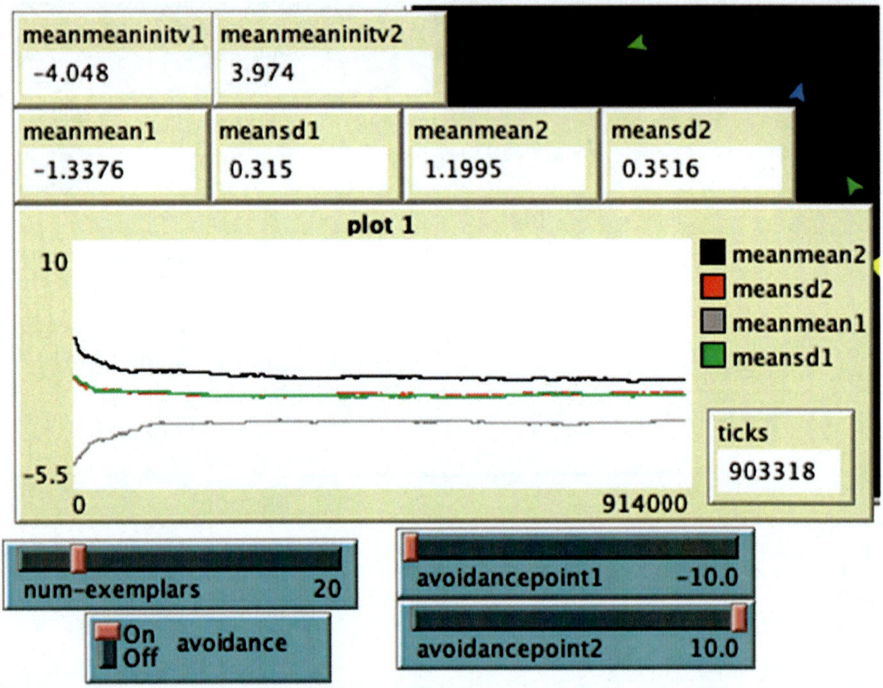

Fig. 15. Model 4: Avoidance and merger: avoidance points of −10 and 10, 37 agents and 20 exemplars each.

PROBE AND ADJUST is a family of reenforcement learning algorithms for learning a continuous value. The algorithm was originally introduced to account for learning continuous moves in market games, such as prices and quantities offered on the market (Kimbrough and Murphy 2009). PROBE AND ADJUST has subsequently been shown to perform well in certain parametric (non-strategic) learning situations as well as game-theoretic cases generally (when continuous plays are present) (Kimbrough 2012), and it has shown effectiveness in complex applications (Kimbrough and Murphy 2013; Haas et al. 2013).

The underlying ideas of the PROBE AND ADJUST family are fairly simple. The learning context is one in which an agent seeks to learn a best or good value of a parameter P. P is typically continuous, but this is not strictly necessary. Further, learning is episodic: the agent repeatedly tries a value of P, receives feedback, and periodically updates its *baseline value* of P. The process is continues indefinitely.

At each episode (think: at each tick of the clock in a simulation), the agent probes by drawing a random deviate based on its current baseline value of P. In the references cited above in this section, probes are drawn from a uniform distribution with range $[P - \delta, P + \delta]$, where δ is a parameter of the algorithm and small with respect to P. Our model 5 in this paper probes with a normal deviate, $x \sim N(P, \sigma)$, where P is the mean and σ is the standard deviation of the normal distribution (and σ is thus a parameter of the algorithm). For each probe the agent makes, a record is kept in one of two stores. If the agent's probe is at or above the mean of the distribution generating it, the probe is stored in the *ProbeHigh* repository; otherwise it is put into the *ProbeLow* repository. Of course, the probe is issued to the agent's environment and has a "real" effect there.

After a number of episodes of probing determined by *EpochLength* (another parameter of the algorithm), the agent adjusts its baseline value, P. In the simplest case, the agent has received feedback from each of its probes and determines whether, on average, it does better probing high or probing low, and it adjusts accordingly. If probing high does better, the agent adjusts its baseline value P by adding ε (a parameter of the model and a small number; we are using 0.5 in model 5); otherwise it subtracts ε. After the two stores are reset to be empty, the agent continues to probe use the new baseline value of P.

Model5simplePandA.nlogo and simplePandA.py, available from the authors, implement PROBE AND ADJUST in Python, which is called from NetLogo and visualized there. The envisioned target system is one of vowel pronunciation on a single dimension (think: formant F1; there is no inherent limitation to a single dimension in our model). We model the tradeoff between ease of pronunciation and ease of understanding. In a multi-vowel system as schematized in Fig. 1, the interactions of these factors may be complex, with intervening social factors such as class and prestige further complexifying the situation. Staying with the aims of our paper, we focus on a simple case, which may serve as a building block example for future modeling.

The agent issues probes (think: vowel utterances on formant F1) and receives feedbacks on (i) proximity to the low-end of the pronounceable range (the `leftbarrier`), with further away being better (ii) proximity to the high-end of the pronounceable range (the `rightbarrier`), with again further away being better, and (iii) proximity to an intermediate `idealpoint` with closer being better. The high- and low-end values represent extremes of difficulty of pronunciation and the `idealpoint` represents ease of understanding. The agent's net score for an utterance is a weighted sum of the individual scores (all in the 0–1 range by using a logistic transform). We use the model parameter `ideal-factor` multiplicatively on the `idealpoint` score, so that values below 1 represent comparative underweighting and values above 1 represent comparative overweighting.

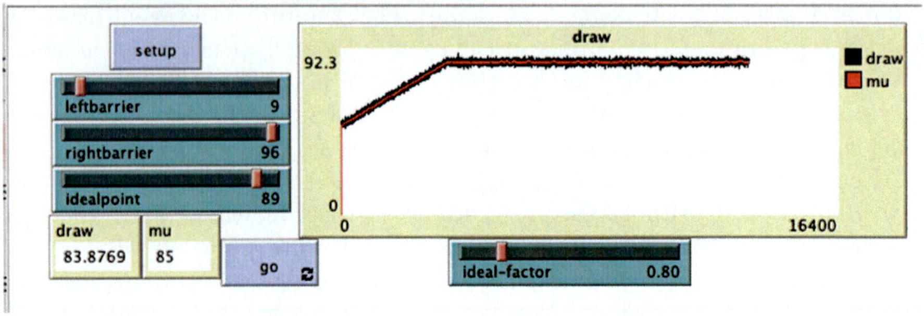

Fig. 16. Model 5: Simple PROBE AND ADJUST.

Figure 16 illustrates a representative run. Here, ease of understanding is slightly depreciated with an `ideal-factor` value of less than 1, 0.8. In consequence, that agent settles on a P value of 85, rather than the 89, which is the `ideal-point` value with respect to ease of understanding.

9 Discussion and Conclusion

In the interests of brevity we note these points arising.

1. We seek to explain, in social and cultural evolution, continuously varying traits that are at the same time discernibly clustered in a common space. Our focus in this paper is notionally on vowel systems, where individual utterances vary continuously in formant space while exhibiting strong association as individual vowels. Many other cultural phenomena share this dual characteristic (projectile points, pottery styles, artistic genres, organizational routines (Nelson and Winter 1982, 2002), and in general any measurable social attribute).
2. We distinguish particulate versus mechanism forms of modeling and explanation for evolution in continuous spaces. In fact this generalizes to discrete spaces., e.g., generate by Poisson or binomial processes. So the fundamental distinction to hand is between particles versus mechanisms/machines as generators of traits and objects of inheritance/persistence.
3. Particulate explanations are in principle possible for our focal traits of interest (continuously varying and clustered), but are prima facie less plausible in most instances. They have proved recalcitrant to identification of the particles, contrary to the view of "memes" adopted by Dawkins (1976) or Blackmore (1999).
4. Acquisition of experience that is used to drive or inhibit a mechanism with continuous output is in principle also a plausible form of explanation. It is what we explore in this introductory study.
5. Our agent representations in models 1–4 are implemented in terms of exemplars, which are discrete. Other conceptual representations are possible, such

as prototypes (Murphy 2002), and collections of parameters as in PROBE AND ADJUST.

6. We have been concerned to identify a number of general building blocks for approaching modeling of this sort. We find it works. We illustrate drift (model 1), directional selection (model 2), and interacting categories to separate (model 3), interacting categories to merge (model 4), and individual learning (model 5).
7. Going forward, the concepts and results of this paper suggest a number of ways to conduct valuable studies of cultural evolution, including the following:
   (a) Descriptive: Identify variables of interest, represent the cultural situation in a landscape or space as in Fig. 1, and track the movement of clusters over time.
   (b) Modeling: Given descriptive accounts of cultural evolution, develop plausible models that can explain what is found in the descriptive accounts. Agent-based models with agents having realistic bounded rationality, as in the models of this paper, appear to be an especially promising methodology in this regard.
   (c) Model calibration: Given modeling accounts, match and fit assumptions (including parameters) of the models to real world data.
   (d) Model testing and validation: Given modeling accounts, use the models to generate testable behavior, both in terms of the model outputs (movements on the landscapes) and in terms of behavior of the model internal mechanisms. Full validation requires testing of both the model's external behavior as well as its internal machinations.

Acknowledgments. We gratefully acknowledge three anonymous reviewers for a number of useful comments and suggestions.

References

Blackmore, S.: The Meme Machine. Oxford University Press, Oxford (1999)

Clark, J.: The Book of Memes. (Independently published)

Clark, R., Kimbrough, S.O.: The spontaneous emergence of language variation from a homogeneous population. In: Proceedings of CSSSA 2015. The Computational Social Science Society of the Americas (2015)

Dawkins, R.: The Selfish Gene. Oxford University Press, Oxford (1976)

Fruehwald, J.: The phonological influence on phonetic change. Ph.D. thesis, University of Pennsylvania, Philadelphia, PA (2013)

Haas, C., Kimbrough, S.O., van Dinther, C.: Strategic learning by e-Service suppliers in service value networks. J. Serv. Res. **16**(3), 259–276 (2013)

Kimbrough, S.O.: Agents, Games, and Evolution: Strategies at Work and Play. CRC Press, Boca Raton (2012)

Kimbrough, S.O., Murphy, F.H.: Learning to collude tacitly on production levels by oligopolistic agents. Comput. Econ. **33**(1), 47–78 (2009). https://doi.org/10.1007/s10614-008-9150-6

Kimbrough, S.O., Murphy, F.H.: Strategic bidding of offer curves: an agent-based approach to exploring supply curve equilibria. Eur. J. Oper. Res. **229**(1), 165–178 (2013). https://doi.org/10.1016/j.ejor.2013.02.006

Labov, W.: Principles of Linguistic Change: Internal Factors. Blackwell Publishers, Oxford (1994)

Labov, W., Rosenfelder, I., Fruehwald, J.: One hundred years of sound change in Philadelphia: linear incrementation, reversal, and reanalysis. Language **89**(1), 30–65 (2013)

Ladefoged, P.: Vowels and Consonants: An Introduction to the Sounds of Language. Blackwell, Malden (2001)

Ladefoged, P.: Representing linguistic phonetic structure. Technical report (2006). https://linguistics.ucla.edu/people/ladefoge/PhoneticStructure.pdf

Levinson, S.C., Dediu, D.: The interplay of genetic and cultural factors in ongoing language evolution. In: Richerson, P.J., Christiansen, M.H. (eds.) Cultural Evolution: Society, Technology, Language, and Religion, pp. 219–232. The MIT Press (2010)

Mesoudi, A.: Cultural Evolution: How Darwinian Theory Can Explain Human Culture & Synthesize the Social Sciences. University of Chicago Press, Chicago (2011)

Murphy, G.L.: The Big Book of Concepts. MIT Press, Cambridge (2002)

Nelson, R.R., Winter, S.G.: An Evolutionary Theory of Economic Change. The Belknap Press of Harvard University Press, Cambridge (1982)

Nelson, R.R., Winter, S.G.: Evolutionary theorizing in economics. J. Econ. Perspect. **16**(2), 23–46 (2002)

Patel, A.D.: Music, Language, and the Brain. Oxford University Press, Oxford (2008)

Vorobeychik, Y., Kimbrough, S., Kunreuther, H.: A framework for computational strategic analysis: applications to iterated interdependent security games. Comput. Econ. **45**, 1–32 (2014). https://doi.org/10.1007/s10614-014-9431-1

Economic Sanctions and Consumer Behavior in Target States: An Agent-Based Model of Boycott Movements

Rena Sung[1(✉)] and Jonghyuk Park[2]

[1] University of Pittsburgh, Pittsburgh, PA 15213, USA
res176@pitt.edu
[2] University of Edinburgh, Edinburgh EH8 9YL, UK
Jay.JH.Park@ed.ac.uk

Abstract. An economic sanction is a foreign policy tool that is used to change the behavior or policies of a target state that the sanctioning state deems unacceptable. Economic sanctions often create negative sentiments among individuals in target states. Such negative sentiments can lead individuals to boycott the sanctioner's branded products in retaliation. We built an agent-based model to examine how individuals start boycotting the sanctioner's branded products in response to economic sanctions and to identify what factors influence the initiation and continuation of a boycott movement in a target state. In the model, we vary four variables: the threat level of the economic sanction, the media's reports on the efficacy of the boycott movement, the utility of the sanctioner's products, and purchase periods. Using these variables, we constructed a baseline model with median values for the four variables. We also created eight scenarios, in which each of the four variables was varied. The results provide new insights into consumer behavior in a target state in response to economic sanctions, as previous studies of economic sanctions focused mainly on public opinion in a target state. We find that the threat level of economic sanctions is crucial in triggering a boycott movement among the first boycotters, and media reports are essential to motivate boycotters to continue the movement. Consumers in a target state are likely to boycott a sanctioner's product if it has high utility and a short purchase period.

Keywords: Economic sanctions · Boycott movements · Target states · Consumer behavior · Agent-based modeling

1 Introduction

Economic sanctions refer to the use of economic tools, such as trade, to change the behavior or policies of other states that the sanctioning states deem offensive [15]. States have increasingly used economic sanctions as a foreign policy tool.

Z. Yang and E. von Briesen (Eds.): CSSSA 2021, SPCOM, pp. 43–55, 2022.
https://doi.org/10.1007/978-3-030-96188-6_4

For example, the EU imposed economic sanctions on Israel to change Israel's tough stance toward Palestine [5]. The US and the EU imposed sanctions on Iran to prevent the development of Iran's nuclear weapons [6]. China imposed economic sanctions on South Korea to change its military policy [16]. Economic sanctions create reactions in target states. Some scholars argue that economic sanctions make the public side with the sanctioner and push their government to change its behavior in favor of the sanctioner [1]. On the other hand, economic sanctions may cause backlash in target states and make the public support their government more [9]. Most studies focus on how the public's opinion changes in response to economic sanctions in target states. In the meantime, some studies have recently examined whether backlash in public opinion leads to backlash in consumer behavior by boycotting the sanctioner's branded products in response to economic sanctions [16].

This paper aims to model how individuals participate in a boycott of a sanctioner's products in a target state as a response to economic sanctions. Economic sanctions are often perceived as a threat to target states. Each sanction differs by threat level; some sanctions are more damaging than others. Individuals in a target state also react differently to sanctions; some people do not take it offensively, but other people react furiously and want to retaliate. We depict how individuals start participating in a boycott and influence others to join them. Consumers also act differently depending on the characteristics of the sanctioner's products. If many people frequently buy the products, then such products are easily targeted and the effect of the boycott on the products is observed in the short term. If products are purchased less frequently due to their characteristics, it will take time to observe the boycott's effect on such products.

The paper comprises four sections. Section 2 introduces related literature and a discussion on boycotts. Section 3 provides an overview of our methodology, including parameters, assumptions, and processes. Section 4 describes the simulation results from our agent-based model of boycotting in response to sanctions. Section 5 concludes the paper.

2 Economic Sanctions, Public Opinion, and Consumer Behavior in Target States

Economic sanctions involve at least two parties: a sanctioning state (sanctioner) and a target state. The economic sanctions have been studied from a sanctioner's perspective, but scholars have recently expanded the literature by incorporating the public's reactions to the sanctions in target states. This is an advancement as a state consists of many individuals whose opinions and behavior matter for policymaking. Such individual-level studies of the consequences of economic sanctions in target states focus on public opinion for which two mechanisms have been put forward. The first mechanism is that those negatively affected by the sanctions push the government to change its behavior in favor of the sanctioner [1,8]. The second mechanism operates through backlash. The public rallies around the flag and supports the target government, intensifying the anger toward the sanctioner [9,11]. In contrast, few studies have examined the behavioral aspect in response to economic sanctions in a target state. For example, Sung and Park [16] incorporated consumer behavior into the economic sanctions literature and examined whether the angry public in a target state would boycott the sanctioner's products in retaliation.

Some literature addresses the public's reactions to international conflict. It is worth noting that the public reacts to the conflict with other states, not economic sanctions in this line of research. For example, Pandya and Venkatesan [14] and Chavis and Leslie [3] found that the sales of "French-sounding" products declined in the US after France opposed the Iraq War. Luo and Zhou [10] examined Chinese consumers' boycott of Japanese products following the territorial dispute between the two countries. These studies show that the public's animosity toward the other state can lead the public to avoid purchasing the other state's products. In this way, the research recorded consumers' reactions to conflict with other states by examining the sales of the products from the states in conflict. However, the studies do not closely depict how a boycott movement is evolved into a nationwide movement.

Boycotting has attributes of collective action. An individual's participation in a boycott has a negligible impact on firms [7]. Thus, an individual can benefit from a successful boycott regardless of whether she or he participates in a boycott. This creates free-rider incentives. In this regard, boycotts are similar to voting, as a single ballot has a negligible impact on election outcomes. However, people still vote [4]. In some sense, boycotting is more puzzling than voting because elections are held on a regular basis, and people can learn from the outcomes of regular elections and have incentives to vote [2]. In contrast, a boycott occurs irregularly, and it requires purely voluntary participation to initiate the boycott.

The free-rider problem of collective acts assumes that individuals are rational and care only about their interests [12]. However, previous studies on collective action have pointed out that people might not opt for a free ride because individuals do not always act out of self-interest [13]. Ostrom [13] also put forward certain conditions that allow collective goals to be achieved without free-rider problems. In a similar spirit, for a boycott to occur, we believe that some factors are essential. In the context of economic sanctions, the seriousness of economic sanctions as a threat, which determines the degree of hostility; the influence of one's peers; and the (perceived) effectiveness of a boycott are all crucial for a successful boycott. By a successful boycott, we mean that the number of people who participate in a boycott is large enough to make a sizable impact on the sales and market share of the company selling the targeted products. Based on the abovementioned crucial factors, we will further develop our model in the next section.

3 Model

We have developed an agent-based model (ABM) with NetLogo to simulate the process of how economic sanctions may (or may not) trigger a boycott move-

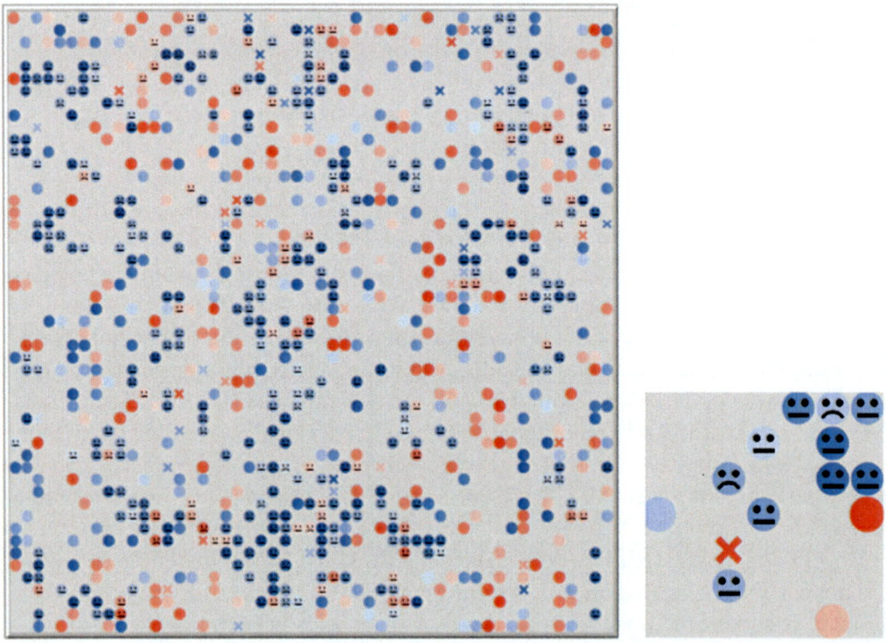

Fig. 1. A snapshot of our sanction-boycott model. Red agents represent consumers who recently purchased the sanctioner's branded product. Blue agents represent consumers who recently purchased a product from a non-sanctioner's brand. Colors of lighter hues mean that the agent has made his last purchase less recently. A neutral-faced agent represents a boycotting consumer, and an unhappy-faced agent represents a consumer who is boycotting and actively propagating the boycotting behavior. An "X" mark represents a consumer who has lost interest and has become "immune" to the idea of boycotting (unless she is persuaded by a neighboring *propagator* to participate again).

ment among citizens in the target state. In our model, consumers have access to two different brands of the same type of product, one of which is from a sanctioning country while the other is from a non-sanctioning country. Inspired by Sung and Park [16], we have identified the *hostility* towards the sanctioner as the driving force of individual-level boycotting behavior. That is, the more threat a consumer perceives from an economic sanction, the more hostile she is likely to become towards the sanctioner, thus, becoming determined to avoid purchasing the sanctioner's branded products in retaliation. Such retaliatory purchasing behavior may spread from person to person and eventually develop into a collective boycott movement. We promote our model as a formal device for studying what makes some boycotts against the sanctioner persist and have lasting effects, while others subside before catching on and becoming a nationwide movement.

Figure 1 is a screenshot of our model visualized during simulation. A population of consumer agents are randomly distributed in a 51×51 grid, where an agent may have up to eight neighbors. Each agent is randomly initialized with attributes named *response to threat*, *patience*, and *propagator*. *Response to threat* is a quantitative measure of an agent's sensitivity to the threat imposed by the sanctioning state, which is sampled from a truncated normal distribution. *Patience* measures how long (in terms of time steps) it takes for an agent to give up the boycotting behavior, unless she gets an "efficacy signal" via mass media reporting that a successful boycott movement is ongoing with a noticeable decline in the market share of the sanctioner's brand. *Propagator* is a Boolean attribute that represents whether the agent will become an active propagator of the boycott movement when she becomes a boycotter.

We maintain a number of environment-level parameters in our boycott model, of which we consider the following to be of most significance and, thus, key variables in our experiments: *threat level*, *media exposure*, *difference in product utility*, and *product purchase period*. *Threat level* encodes the extent to which the economic sanction is threatening to the target state as a whole. This environment-level parameter is taken together with the individual-level attribute of *response to threat* to yield the agent's hostility towards the sanctioner as its state variable. *Media exposure* quantifies how frequently the consumers can hear from the media about the development of the ongoing boycott movement, which helps them sustain their momentum by providing a sense of efficacy. By *difference in product utility*, we parameterize the competitive edge of the sanctioner's product over the non-sanctioner's product in terms of utility, which affects how likely it is for a consumer in a natural state to choose the former product over the latter when making a purchasing decision. Finally, by *product purchase period*, we control whether the product of interest is of such type that is consumed more frequently (typically with a lower price), or less so (with a higher price).

In a single run of simulation, we first randomly initialize the consumer population according to some provided population density in the 51×51 grid, so that each patch on the grid may be either unoccupied or have a single consumer agent. After we sample the agent attribute values mentioned above for each agent, we randomly assign each agent a binary value indicating whether she has made her latest purchase from the sanctioner's or the non-sanctioner's brand (represented by colors in Fig. 1). The value is sampled from the probability computed from the product utility difference parameter; this exact probability will also serve to sample the purchase decision made by each non-boycotting consumer during simulation. To wrap up the initialization process, the hostility value is computed for each agent from the threat level parameter and the individual threat response attribute. Agents with hostility values higher than a pre-determined threshold voluntarily become boycotters. While we assume in this paper the threat level is constant across time steps and, thus, put the hostility computation in the initialization phase, we can easily move the computation step into the main simulation loop if we want to extend the model by treating the threat level as a dynamic variable.

After initialization, our model follows the flow depicted in Fig. 2. Specifically, our model executes the following subprocesses for each consumer agent in the given order at each simulation time step:

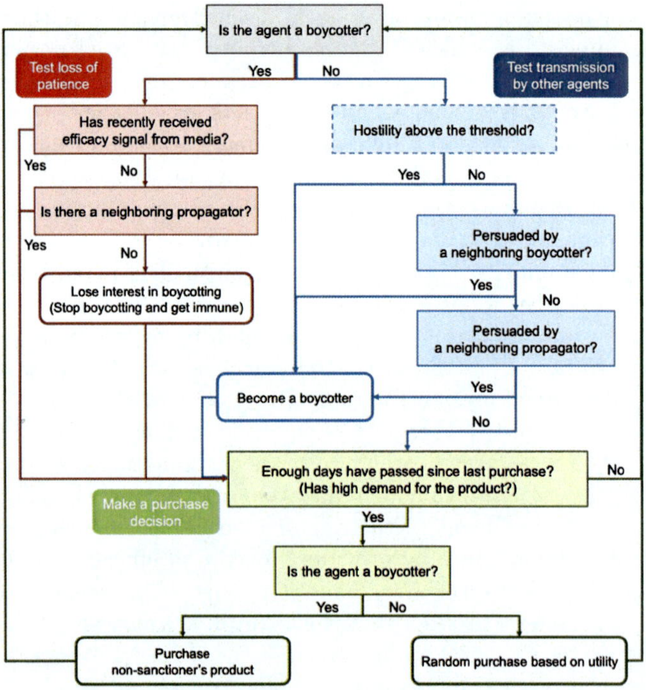

Fig. 2. Flow of model

- **Test loss of patience**: Boycotting is a costly choice by definition. To express discontent, boycotting consumers give up their preferred option that would provide higher utility than other alternatives. Unless boycotters are rewarded with some signal that their sacrifices are actually making an impact, it is natural to expect that they will eventually lose motivation for maintaining their suboptimal consumption behavior. In most cases, the primary source of such efficacy feedback is communication from the mass media; for example, news articles reporting that the sanctioner brand has lost a notable portion of market share since the outset of the boycott movement. We model this mechanism for each boycotter by testing whether more time steps have passed than she can wait before getting a new efficacy signal[1] from the media (i.e. *patience* attribute), which may broadcast at each time step with a probability defined by *media exposure*. When this happens, the agent loses patience and stops boycotting, becoming immune to the idea of boycotting the product. The immune agent may resume boycotting if they are persuaded by the propagators later.
- **Test transmission by neighborhood effect**: Neighbor influence is a critical part of the model. Not all consumers decide to boycott on a voluntary basis; some decide to do so because it is a popular trend among the people

[1] Refer to the Appendix for the precise definition of *efficacy signal* in our model.

around them. It would be reasonable to expect that consumers who have a higher chance of interacting with boycotting consumers are more likely to be influenced by them to become a boycotter as well. This could be due to peer pressure, a higher likelihood to make purchase decisions together, and so on. In this regard, our model is somewhat analogous to an epidemic model where a behavioral meme propagates among a population through social interactions. In our model, a non-boycotting consumer agent interacts with its neighbors in each time step. For each interaction with a boycotting neighbor, there is a (parameterized) probability that the non-participating agent will start boycotting as well.

- **Test transmission by active propagator**: It is not uncommon to see that some boycotting consumers are more active in propagating their mission to convert non-boycotters into boycotters. This might be due to their personality, exposure to a campaign that urges participants to disseminate boycotting behavior, and so on. We model the effect of the presence of such ardent persuaders separately from the neighborhood effect explained above. In addition to having a higher chance of transmission, we model this effect to be qualitatively distinct from the neighbor effect by allowing them to persuade those who become immune to boycotting to boycott again.
- **Make a purchase decision**: At the end of each time step, each consumer who has not purchased a product for over a designated period, makes a purchase decision. In the absence of any threat posed, the purchase decisions made by consumer agents will be sampled according to a probability value that gives products with higher utility a greater chance to be selected. However, a boycotting consumer will ignore utility differences and buy products only from the non-sanctioner brand. Naturally, the consequence of a boycott movement is heavily affected by the characteristics of products targeted for a boycott. In this study, we would like to investigate how the development of a boycott movement may vary according to two parameterized characteristics of the competing products: *difference in product utility* and *product purchase period*. The implications of different product characteristics can be explained in terms of initiation and continuation of a boycott movement. Popular branded products generally give consumers high utility and have large market shares. Accordingly, a sanctioner's popular branded products can be easily targeted for a boycott as people are well aware that the product is the sanctioner's brand. It is obviously costly for consumers to give up on their preferred option, but it is the product that can make the biggest impact on the sanctioner's sales, thus leading consumers to start a boycott. Moreover, a large decline of the product with the highest market share will keep boycotters motivated to continue the boycott because successful boycotts of popular products send out stronger efficacy signals to consumers. Different purchase periods also affect how quickly the consequence of a boycott movement is reflected in the market. People tend to target products that are more accessible to them in terms of the purchase period. For this reason, boycotts of consumer goods with shorter purchase cycles (e.g. beer and apparel) are likely to be targeted first and generate more immediate responses.

– **Update global statistics**: After all consumer agents have updated their state variables and made purchase decisions (if applicable), the time step is completed by making necessary updates to global variables. We are primarily concerned with two outcome statistics to be reported, namely *market share* and *boycotting consumer ratio*. In our model, the market share of the sanctioner's product at a time step is defined as the proportion of consumers whose most recent purchase was from the sanctioner's brand. For example, the number of red agents divided by the total population in Fig. 1 would be the market share of the sanctioner's product. The definition of boycotting consumer ratio is straightforward and self-explanatory.

We do not, of course, aim to implement every single minor aspect of boycott movements provoked by economic sanctions that can be observed in real-world cases, and our model is an abstraction with several simplifying assumptions. We assume that there is no artificial intervention to the market except the threat initially posed by the sanction. We also assume that the possibility of media reporting on boycott-related news stays constant over time. The threat level remains constant across time steps, and the relative utilities of the sanctioner's and non-sanctioner's products remain constant as well. We only consider two products (sanctioner's and non-sanctioner's) in the whole market of the product type. Finally, we assume boycotters will never surrender as long as they receive the signal from the media that their retaliatory actions are taking effect.

4 Results

We examine four main variables that are essential to understanding how individuals start and continue boycotting in retaliation to economic sanctions: the threat level, which determines the degree of hostility toward the sanctioner; the frequency of the media's reports on the efficacy of a boycott (media exposure); the product utility of the sanctioner's product; and the product purchase period. Our baseline model is set at median values for these four factors. In addition to the baseline model, we have eight scenarios, in which we vary one of the four variables while holding the remaining three variables at baseline values.

A simulation for each scenario was run 100 times to obtain robust and stable outcomes from the model simulation, which has a stochastic nature. We simulated each scenario for 400 time steps. For each scenario, we wanted to know how long a boycott lasts after its inception and whether the boycott is likely to be permanent, that is, people continue to boycott the sanctioner's products to the point that not purchasing the sanctioner's products becomes a habit for many individuals. We were also interested in the boycotters, particularly those who participate in the boycott from the outset (time step one). The highest percentage of boycotters was also captured to see the efficacy of the boycott. We also recorded the changes in the market share of the sanctioner's product. It was calculated as the difference between the initial market share of the product before the economic sanction was imposed and the lowest market share of the product

Table 1. Summary of results by scenario

Main factors/Scenarios	Baseline	High threat	Low threat	High media	Low media	High utility	Low utility	Frequent Purchase	Infrequent purchase
	(1)	(2)	(3)	(4)	(5)	(6)	(7)	(8)	(9)
Threat	0.5	0.7	0.3	–	–	–	–	–	–
Media reporting	0.5	–	–	0.9	0.1	–	–	–	–
Product utility	0	–	–	–	–	2	−2	–	–
Purchase period	100	–	–	–	–	–	–	50	300
Results									
Number of time steps boycotting	267.04	397.68	105.59	295.35	178.71	336.29	226.25	258.32	114.24
Standard deviation	83.07	13.20	45.47	67.63	55.28	94.63	73.34	56.50	41.13
Boycott likely to be permanent?	No	Yes	No	No	No	No	No	No	No
Max percentage of boycotters	0.38	0.78	0.04	0.43	0.26	0.41	0.37	0.46	0.23
Standard deviation	0.13	0.05	0.02	0.13	0.06	0.12	0.13	0.09	0.04
Δ in market share of product	0.18	0.39	0.04	0.21	0.10	0.35	0.04	0.23	0.04
Standard deviation	0.08	0.02	0.02	0.07	0.05	0.13	0.02	0.05	0.03
The percentage of first boycotters	0.16	0.50	0.02	0.16	0.16	0.16	0.16	0.16	0.16
Standard deviation	0.01	0.02	0.01	0.01	0.01	0.01	0.01	0.01	0.01

during the boycott. A high value means that individuals drastically cut down the purchase of the sanctioner's product. Table 1 describes the summary of results by scenario with mean values and standard deviations from 100 simulations.

For the baseline model (which has median threat level and media reporting values, a utility difference of zero, and a purchase period of 100 time steps), we observed that the boycott lasts for 267.04 time steps on average. Here we assume that seven time steps are equivalent to one day in the real world. The first boycotters comprised 16% of the population at time step one, and the highest boycott consumer percentage was 38%. The market share of the sanctioner's product declined by 18%.

Regarding the threat level, we can see how a high threat level is likely to lead to a permanent boycott. Note that other values are set at the same values as the baseline model. When the threat level is high, meaning that many individuals are hostile toward the sanctioner, the percentage of those who initiate the boycott becomes substantially higher than that of the baseline model. Similarly, the market share of the sanctioner's product declines substantially, compared to the baseline model. In contrast, when the threat level is low, the percentage of the first boycotters is substantially lower than the baseline model, and others do not join the boycott. The market share barely declines in this case.

Regarding the frequency of media reporting on the efficacy of the boycott movement, when the frequency of media reports is high, the boycott lasts longer than the baseline model on average. This tells us that frequent feedback of how the boycott is progressing provides incentives to boycott participants to continue the movement. Accordingly, the market share of the sanctioner's product declines more than the baseline model. We find a contrasting pattern when the frequency of media reports is low. Compared to the baseline model, the boycott lasts shorter with a smaller decline in the market share. The maximum percentage of the boycotters is also smaller than the baseline model.

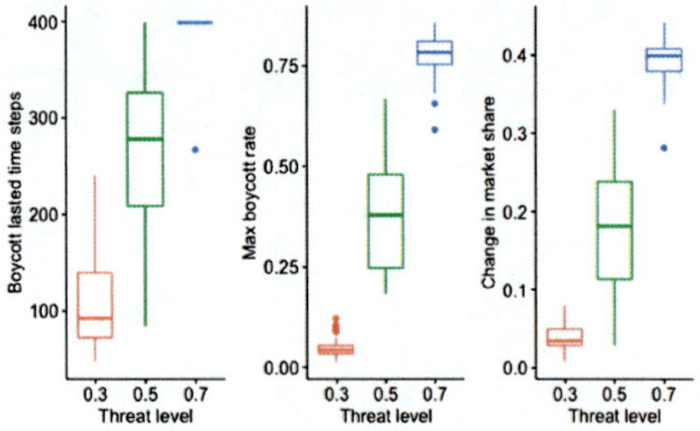

(a) Threat

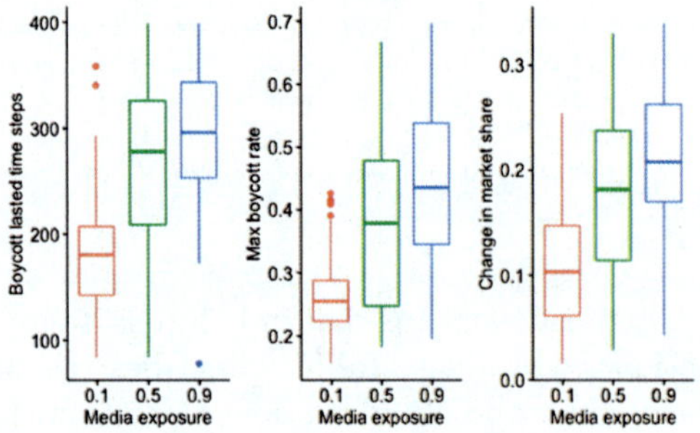

(b) Media exposure

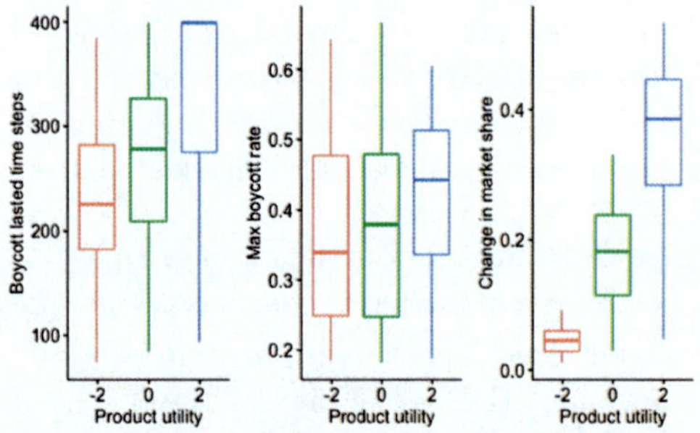

(c) Product utility

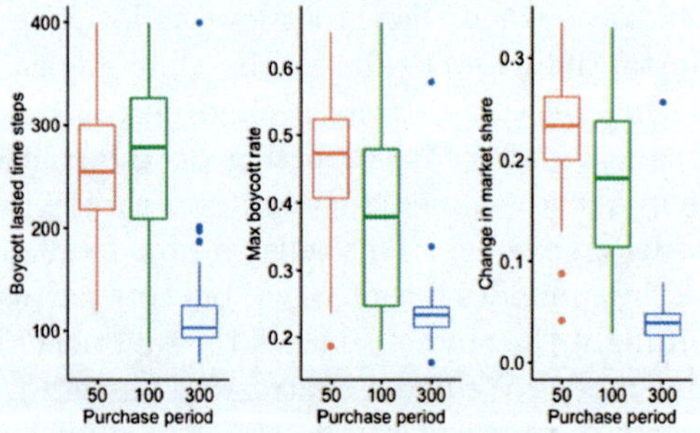

(d) Purchase period

Fig. 3. Summary of outcomes by variable

Consumers prefer some products over others based on utility calculations. The decline of the market share of a product with high utility is large, compared to when there is no utility difference in products between the sanctioner and the non-sanctioner. In contrast, the market share of a product with low utility declines in a small magnitude. It implies that a sanctioner's product that provides high utility takes a large portion of the market share and is well recognized as the sanctioner's brand. Individuals sacrifice the joy of buying such a popular product when they become hostile towards the sanctioner. In other words, products that are not popular to begin with are less likely to be targeted. Although individuals can still boycott such a product, the impact is relatively small.

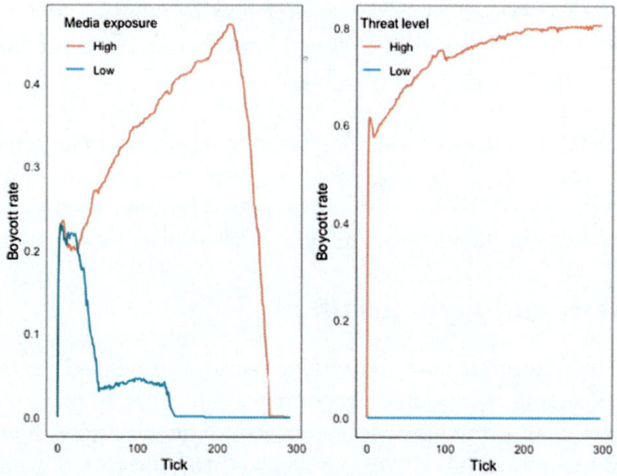

(a) Boycott rates

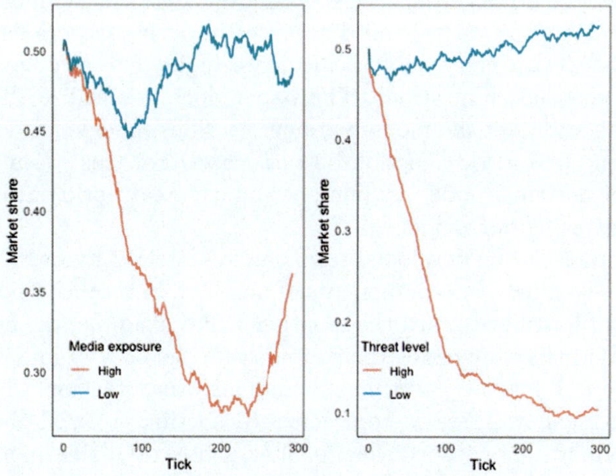

(b) Market share

Fig. 4. Market share and boycott rate in response to economic sanctions

Consumers also purchase products at varying frequencies. In our model, the market share of a sanctioner's product that individuals buy frequently declines more than the baseline model, whereas that of a product with a low purchase frequency declines only slightly. This implies that the products with a short purchase period are more likely to be targeted for a boycott, as they come to consumers' minds more easily and frequently than those products that consumers purchase less frequently. Moreover, the hostility might go away by the time consumers decide to purchase products with a long purchase period. Figure 3 presents the box plots of the summary outputs by variable. We also run the Wilcoxon test and found that, for all variables, the outputs of the eight scenarios are significantly different from those of the baseline model, respectively.

Figure 4 plots the trends of boycott rates and the market share of the sanctioner's product when the level of a threat and media exposure vary from low to high, while holding other attributes baseline values. The key finding from our simulation results is that a high threat level increases hostility toward the sanctioner, which in turn, increases the boycott participation rates from the outset of the economic sanction. This leads to more participation from non-boycotters, increasing the efficacy of the boycott movement. Frequent media reports on the boycott's efficacy further lead to ongoing boycott participation from boycotters.

5 Discussion and Conclusion

In this paper, we examine how individuals become involved in boycotting in retaliation to economic sanctions. We believe that this is an advancement of the subject matter as previous studies on economic sanctions focus on public opinion only. By using an ABM, we can see how the boycott movement spreads among individuals and how the main factors influence individual purchasing decisions. We find that a high threat level that creates hostility among individuals is essential to triggering a boycott movement. The frequency of media reports, the utility of products, and differences in purchase periods also matter in promoting boycott activities and achieving a highly effective boycott.

Our paper also has limitations. The paper does not utilize the real world data in deriving values of the model parameters. This model can be made more applicable to the real world sanction-boycott cases. For this, we plan to devise mechanisms for deriving model parameters from the real world data and explain real world cases using our model.

Further studies can be developed from this research in two ways. One direction is to delve deep into the components of hostility. Several components determine the level of hostility toward the sanctioner. For example, the hostility components of individuals can include their support for the policy that the sanctioner challenges, support for the government, and historical tension with the sanctioner. Examining how these factors generate hostility toward the sanctioner will provide a more fine-grained view of how a boycott gains momentum and spreads throughout the country.

The other direction of the research is to further examine product types that are more likely to be targeted for a boycott by consumers in a target state. In this paper, we only look at two product characteristics: utility level and purchase period. We can explore product types in more detail, using the real-world data. For example, even among products with a high level of utility and thus high popularity, some are more likely to be targeted for a boycott than others. It will be interesting to explore how individuals participate in a boycott based on product types.

References

1. Allen, S.H.: The domestic political costs of economic sanctions. J. Conflict Resolut. **52**(6), 916–944 (2008)
2. Bendor, J., Diermeier, D., Siegel, D.A., Ting, M.: A Behavioral Theory of Elections. Princeton University Press, Princeton (2011)
3. Chavis, L., Leslie, P.: Consumer boycotts: the impact of the Iraq war on French wine sales in the us. QME **7**(1), 37–67 (2009). https://doi.org/10.1007/s11129-008-9043-y
4. Downs, A.: An economic theory of political action in a democracy. J. Polit. Econ. **65**(2), 135–150 (1957)
5. Grossman, G., Manekin, D., Margalit, Y.: How sanctions affect public opinion in target countries: experimental evidence from Israel. Comp. Pol. Stud. **51**(14), 1823–1857 (2018)
6. Haidar, J.I.: Sanctions and trade diversion: exporter-level evidence from IRAN. In: ERF Annual Conference: Economic Development and Social Justice (2014)
7. John, A., Klein, J.: The boycott puzzle: consumer motivations for purchase sacrifice. Manag. Sci. **49**(9), 1196–1209 (2003)
8. Kaempfer, W.H., Lowenberg, A.D., Mertens, W.: International economic sanctions against a dictator. Econ. Polit. **16**(1), 29–51 (2004)
9. Kam, C.D., Ramos, J.M.: Joining and leaving the rally: understanding the surge and decline in presidential approval following 9/11. Public Opin. Q. **72**(4), 619–650 (2008)
10. Luo, Z., Zhou, Y.: Decomposing the effects of consumer boycotts: evidence from the anti-Japanese demonstration in china. Empir. Econ. **58**(6), 2615–2634 (2020). https://doi.org/10.1007/s00181-019-01650-3
11. Mueller, J.E.: Presidential popularity from Truman to Johnson. Am. Polit. Sci. Rev. **64**(1), 18–34 (1970)
12. Olson, M.: The Logic of Collective Action: Public Goods and the Theory of Groups. Harvard Univ. Press, Cambridge, Mass. (1965). http://en.wikipedia.org/wiki/The_Logic_of_Collective_Action
13. Ostrom, E.: Governing the Commons: The Evolution of Institutions for Collective Action. Cambridge University Press, Cambridge (1990)
14. Pandya, S.S., Venkatesan, R.: French roast: consumer response to international conflict-evidence from supermarket scanner data. Rev. Econ. Stat. **98**(1), 42–56 (2016)
15. Pape, R.A.: Why economic sanctions do not work. Int. Secur. **22**(2), 90–136 (1997)
16. Sung, R., Park, J.: How do economic sanctions affect public opinion and consumer behavior in target states?: Evidence from China's economic sanctions on South Korea. Forthcoming in International Studies Quarterly

Check for updates

Scheduler Dependencies in Agent-Based Models: A Case-Study Using a Contagion Model

Srikanth P. Mudigonda[1(✉)], Santiago Núñez-Corrales[2], Rajesh Venkatachalapathy[3], and Jeffrey Graham[4]

[1] School for Professional Studies, Saint Louis University, St. Louis, MO, USA
srikanth.mudigonda@slu.edu
[2] NCSA, University of Illinois Urbana-Champaign, Urbana, IL, USA
nunezco2@illinois.edu
[3] Systems Science Graduate Program, Portland State University, Portland OR, USA
venkatr@pdx.edu
[4] Department of Mathematical Sciences, Susquehanna University, Selinsgrove, PA, USA
graham@susqu.edu

Abstract. Scheduler dependencies in Agent-Based Models (ABM) simulations are known to exist, but often remain sidelined in practice, and continue to be theoretically and technically opaque. Understanding whether the differences introduced by scheduling choices yield significant deviations from both analytic results and phenomenological observation can provide insights on whether those discrepancies may lead to meaningful differences of scientific interpretation or decision making. Here, we demonstrate their presence and non-trivial nature in a variant of the standard agent-based SIRD contagion model, one in which an agent's local decisions are informed by both their local environment and the consensus of their spatially non-local social network, where event priorities are computed from the network centrality of each agent. Preliminary simulation outcomes suggest that priority scheduling introduces resonant stochastic fluctuations modulated by social mimicry.

1 Introduction

Fidelity to natural phenomena and processes and mechanisms that generate them is an aspirational ideal for modeling exercises in the ABM world. Modeling processes and mechanisms at a single level is hard; it is an order of magnitude or more harder for models that try to explain the macro from the micro, as there is potentially an infinite number of generative micro models that give rise to the same macro phenomenon of interest. Hence, finding the most adequate parameters and their mechanistic interpretation may not be either straightforward or

S. P. Mudigonda, S. Núñez-Corrales, R. Venkatachalapathy and J. Graham—SPEC SIG.

Z. Yang and E. von Briesen (Eds.): CSSSA 2021, SPCOM, pp. 56–70, 2022.
https://doi.org/10.1007/978-3-030-96188-6_5

feasible. Further, it relies on heuristic model selection criteria, analogs of which are well-known in inferential statistics. One such heuristic, adopted perhaps for the sake of expediency, may be the assumption that once an ABM is specified, the scheduler (which manages the sequence in which agents are activated) used by the modeling platform might not make a difference to the phenomena generated. Using this heuristic allows the modeler to go with the default agent-activation scheme, e.g., random activation. In the work reported here, we demonstrate how the results, in form of the trajectory followed by the outcome of interest, can vary as a function of the scheduling approach used for agent activation, specifically in an agent-based contagion model. Here, we provide evidence of presence and significance of scheduler dependencies, substantiating a need for a scheduler taxonomy (Weimer et al. 2019).

After an elaboration on why ABM schedulers ought to be part of the model under scrutiny in the next section, we describe in the subsequent section a coupled-contagion model, where scheduler dependencies are shown to lead to differences in model output. Finally, we conclude by discussing future research directions.

2 Fidelity to Social Processes and Relevant ABM Mechanisms

The questions and concerns raised here and elsewhere (Núñez Corrales et al. 2020) are at once new and old: new because of their framing within analytical sociology and social mechanisms (Hedström et al. 2009); old because these questions have been, and continue to be, discussed by practitioners studying verification and validation of ABM (Klügl 2008; Cooley and Solano 2011), and by philosophers studying epistemological foundations of computer simulations (Lake 2015; Epstein 1999; Winsberg 2019; Frey and Šešelja 2018). The relatively unsettled nature of discussions might have to do with the contraposition between the underpinnings of conventional models of science - equations-based models.

Scheduler dependencies, the topic of interest herer, has analogs in numerical implementation of equation-based models, in computationally implemented numerical techniques for solving ordinary differential equations (ODEs). Typically, when an ODE is solved numerically, it makes use of some time discretization techniques, for example, the Runge-Kutta technique. However, while this works for a wide variety of ODE models, the technique fails for other classes called stiff-ODEs; for these cases, varying step size techniques are required to stabilize the solvers. In practice, solvers come with options to use different recipes, some of them for solving stiff ODEs. Simulation workflows involve comparison of solutions across a few different solvers. The argument here is that the schedulers are analogs of constant step-size-varying step-size recipes for ODEs, and just like in the ODE case, one has to test to determine whether different schedulers produce different results.

In the absence of a rigorous formalism for studying fidelity of ABM models to (social) reality, a reasonable baseline formalism could serve as a starting point.

One such candidate is the class of interacting particle systems (IPS) (Aldous 2013). For such classes, which include models of opinion formation, contagion, and even cellular automata, particles with reasonable microscopic properties reduce to macroscopic ODEs and partial differential equations that are well-known: susceptible, infected, recovered (SIR) models, reaction-diffusion equations, predator-prey models, pattern formation models, etc. When agents have memory, there is concurrency, with multiple interactions happening independently at different timescales. Additionally, multi-agent interactions are present and the resultant macrobehavior does not reduce to conventional equation-based models: generative models like ABMs are more expressive than mathematical models. These difficulties have led in part to the search for a generalized theory of interactions for complex multiscale stochastic systems (Núñez-Corrales and Jakobsson 2020) of which ABMs are an example.

Most ABMs specified in practice do not have the complex characteristics listed above; they are simpler than they could be because one expects that models with incrementally more generality suffice, even if they are systematically misspecified, because the model classes in use are good enough approximations of a truer, possibly more refined model. But what if these models are insufficient? Then, the model output of the more refined ABM, and the good-enough or coarser-grained ABM, are likely to differ. This has been demonstrated in several systems[1]. Asynchrony, for instnace, can lead to ordering artifacts in simulation outputs when implemented in discrete-time. When using an ABM platform, if a modeler uses the default (sometimes called *base*) scheduler, an unintended consequence may involve breakage of the finer temporal details of the consequences of asynchrony. Such instances may be seen in opinion dynamics models (Weimer et al. 2019); in cellular automata models (Page 1997; Cornforth et al. 2005; Bouré et al. 2012; Fates 2013; Reia and Kinouchi 2015); in ecological models (Caron-Lormier et al. 2008); in multi-agent systems (Fatès and Chevrier 2010; Bonnell et al. 2016); and in epidemic models (Shu et al. 2016). Comer (Comer 2014) noted the existence of differences in outcomes of three types of social models when the activation function in the scheduler was changed; our work expands on his findings.

We implemented asynchrony by coupling the scheduler and the model, based on the observation that true event reorderings are different from mere reorderings in the labeling of the agents. A prioritized random scheduler operates by subdividing each simulation timestep and performing computations that couple model factors determining the priority of events. We assume a) that the phenomenon being modeled manifests one scale of local action operating under a local clock that can be discretized, b) that such a scale can be divided into instants driven by clocks operating at a higher rate characteristic of other local or non-local processes, and c) that the faster processes can mechanistically reconfigure the causal paths of distinguishable events in the slower scale. Practically, the coupling between epidemic and infodemic scales in COVID-19 (Gallotti et al. 2020) provides an insight into how social networks acting orders of magnitude faster

[1] See literature surveyed in (Núñez Corrales et al. 2020).

than biological contagion can reshape a pandemic. Our implementation differs from the staged activation scheduler in the Python-based Mesa ABM package in this sense (Kazil et al. 2020).

Next, we report on our study of one variant of a single individual-based contagion model that couples local epidemic interactions to non-local social information transmission. We demonstrate scheduler dependencies that arise while simulating this model, and point out the quirky and partially unpredictable nature of its presence, where it is sometimes amplified and at other times suppressed.

3 Case Study

Our model is a stylized model designed for exploring ordering or tiered-information gathering and updating processes. This is simulated with different agent-activation schedules: random activation and prioritized random activation. Additionally, our model aims to be representative in general of classes of phenomena where epidemic-like contagion interacts with information transmission. The spirit of this case study is identical to papers such as (Caron-Lormier et al. 2008): propose two different state-update schedulers; and probe for differences in simulation results. While Caron-Lormier et al.'s ecological model uses concurrency-respecting snapshots of energy states and compares it with an update-as-you-go scheduler, our model stratifies agents based on the social ranking and updates them with a ranking-dependent priority scheduling.

3.1 Description of the Model

Our work builds on a standard susceptible-infected-recovered-deceased (SIRD) agent-based compartmental model on a grid world of fixed size, and a small-world (Watts and Strogatz 1998)[2] social network. Agents are persistent random-walkers: they leave a given site probabilistically; they make random jumps of a given length from site to site within the grid. Movement across the grid using single jumps assumes a Moore neighborhood. When two agents meet on a grid point, biological contagion is possible. When an infected agent comes in contact with one or more susceptible agents, each susceptible agent is infected probabilistically; once infected an agent continues to be infected for a period of time, either recovering, or dying (with a small probability); once recovered, an agent remains in that state. At every tick of the global clock, one of two timescales in the model, in the simulation, agents move, perform updates according to both local conditions and their internal state. A state transition diagram that depicts the SIRD compartmental model is shown in Fig. 1 on the following page.

[2] This citation is pertinent as it is the basis of the small-world network generation implementation used in `LightGraphs` (Bromberger et al. 2017). The work builds on a standard susceptible-infected-recovered-deceased (SIRD) agent-based compartmental model on a grid world of fixed size, and a small-world package used in simulations.

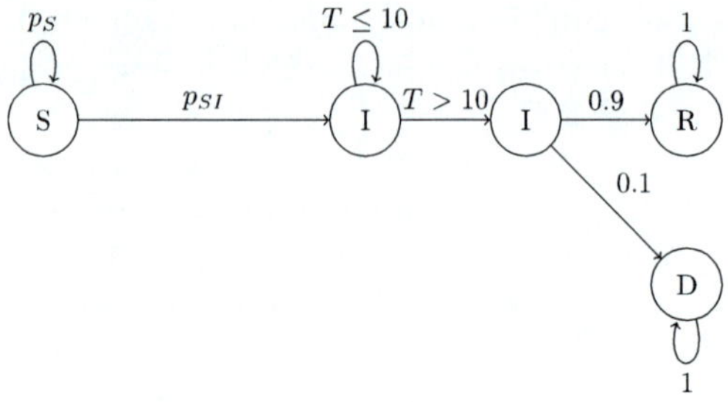

Fig. 1. State diagram of the SIRD compartmental model used to perform scheduler experiments.

A second timescale for the decision process driven by social factors was implemented via the prioritized random scheduler. Agents are earmarked, and the specific earmark depends on conditions that can be either fixed or dynamically computed. Social contagion, i.e., social mimicry of mobility patterns is mediated by an underlying small-world social network, which also stratifies the population into three categories, based on their degree centralities. The agents' group-based earmarks are used by the prioritized random scheduler. Agents with higher centrality or status move and revise their social behavior ahead of agents with lower status. An agent's mobility is probabilistically influenced by peer social behavior resulting in social mimicry[3]. Social mimicry appears to be fundamental across social phenomena (Ashton-James et al. 2007), can produce strong nonlinear dynamics (Marinoff 2001), and has shown to be of striking relevance during the ongoing pandemic (Mistur et al. 2020).

The process of computing the earmarks and performing the staged activation of agents occurs within one step within the coarser time scale. State update events within agents sharing the same earmark are fully randomized. Table 1 describes the details corresponding to the parameters of our model.

[3] Social mimicry: in the condition when social mimicry is set to true, in a given time step each agent that is activated (whether through random-activation or prioritized random-activation), chooses to mimic the behavior of move or not move exhibited by a majority of its peer agents in the previous timestep, for $timesteps > 1$. At time $step = 1$, the agent, based on a probability (drawn randomly from a $Uniform[0, 1.0]$ distribution, in comparison with its activation attribute, chooses to either move or not move. An agent's peer agents are all those agents whose current status is not deceased. The majority is a proportion, whose value is set via a model parameter that is constant for a given simulation run.

Table 1. List of model parameters

Model parameter	Description
`Random seed`	PRNG seed
`Shuffle`	Indicates whether the activation of agents at each time step is random or prioritized random
`Max_agents`	Population size of agents
`Max_dim`	Size of the square grid. Set to 40 across runs
`Inf_rate`	Proportion of agents at the beginning of a simulation run that are infected. Needed because the simulation is a closed system, without any infected agents entering the grid exogenously
`T_infection`	Threshold of immunity below which an agent that is in a susceptible state changes to an infected state upon coming into contact with an infected agent (p_{SI})
`T_change_inf_state`	Number of days after which an infected agent's state changes to either recover or deceased
`Health_red_pct`	Percentage of reduction in the immunity level of an agent that is infected, at each time step
`Degree`	Degree parameter, at the population level for generating a Strogatz-Watts network (K)
`Beta`	The beta parameter, at the population level for generating a Strogatz-Watts network (β)
`Maj_pct`	The proportion of agents in a focal agent's social network whose actions (of moving/not moving) are mimicked in a current time step by the focal agent
`Single_step`	Whether the agents in a simulation run can move only up to one step or not (in which case they can move up to n steps, as determined by their mobility attribute's value)
`Mimicry`	Whether agents in a given simulation run mimic other agent's behavior of moving/not moving or not. When not, there is no social network influenced movement behavior
`Mobility`	Agent-level - maximum distance an agent can move in a timestep, if agent is in multi-step mode
`Health`	Agent-level - the initial level of health
`Priority`	Agent-level - the priority level to which an agent belongs. Initially, all agents are at the level "`low`." At each timestep, their priority is determined by their relative ranking on the basis of their degree centrality: "`high`" priority to the top 1/3rd (in descending order) of degree centrality, "`med`", belong to the middle 1/3rd, and "`low`" to the bottom 1/3rd
`Timesteps`	Total number of timesteps in each simulation run - constant across runs, set to 100

3.2 Experimental Design

The design is a full-factorial approach that systematically sampled 9 parameters. Table 2 on page 5 describes the factors for the design of experiments[4]. For the scheduler type, `pr_random` refers to the prioritized random scheduler, while `random` refers to the usual random activation scheduler present across all ABM packages. A total of 20 runs were performed per each parameter set. Our ABM model was developed and executed in `Julia`(Bezanson et al. 2017) (v1.6.1) and data were analyzed in `R` (v4.1.0) on a laptop with Intel i7 chip with 4 logical cores and 16 GB RAM, running Kubuntu OS v. 21.04. The simulation code used in this experiment, along with data and the code for analyzing the data will be made accessible through GitHub.

Table 2. List of factors

Model parameter	Possible values
Scheduler Type	`pr_random` or `random`
Population Size	`pop.size` = `100` or `1000`
Social Mimicry	`off` or `off`: if `on`, then the agents mimic the behavior if 2/3rds of its neighbors perform the same behavior
Small-world parameters	`degree` = `3` or `9` and `beta` = `0.3` or `0.9`
Death after infection	`0` or `0.9`
Agent Step Size	`Single_step` or `Multistep`: if `Multistep`, then they are further divided into two mobility types
Agent Mobility Types	`homogeneity` = `off` and `on`: `off` conditions imply choosing mobility from a discrete uniform distribution `range[1.,max.dim]`, with `max.dim` = `40`; `on` conditions mean two sub-population with stepsize = `2` or `10`

3.3 Results: Social Mimicry Modulates Scheduler-Dependent Differences

In the following figures, the proportion of agents in the infected (green), recovered (red) and deceased (blue) compartments is displayed using a logarithmic scale in all plots. We began by understanding progressively how the introduction of factors impacts simulation outcomes, with a particular focus on social ones. In the current scenario, the SIRD model produces an increase of infected agents followed by mostly their recovery, with a number of deceased agents around an order of magnitude smaller. Additionally, adding more agents should increase time to convergence if all SIRD parameters remain the same. While results are

[4] NOLH was used initially, but we switched to factorial design to delineate specific differences in the effects of the two schedulers.

reported only for $p_{SI} = 0.05$, we have also carried out experiments with $p_{SI} = 0.15$ with similar outcomes.

Figure 2 displays the outcome when two different schedulers are used (Fig. 2a: random, Fig. 2b: prioritized random). Time to convergence is directly proportional to population size using both schedulers without social mimicry as well as the proportion of agents that become infected. Also, single-step mobility appears to yield a lower convergence point than multistep mobility. This step allowed us to ensure that our implementation of the priority scheduler contains the random scheduler as a special case by turning social mimicry off. To explore more fully the effect of the prioritized random scheduler, we turned to understanding the effect of single-step vs. multistep in connection to β (Fig. 3). Adding social mimicry changed the SIRD model by adding oscillations embedded in the contagion curve of the infected compartment. With $K = 3$, the difference between $\beta = 0.3$ and $\beta = 0.9$ became negligible when single-step is used, but not for multistep. In addition, an inspection of the case $K = 9$ suggests that increasing the network degree has the potential to yield more models with local minima (two in Fig. 3b) than when the network degree is smaller (one in Fig. 3b). This wavelike behavior is not unexpected since reaching different levels of consensus depending on specific conditions depends on spatial and social models of contagion.

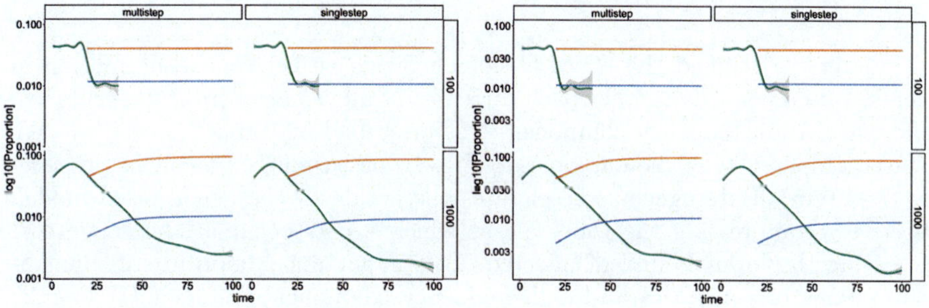

(a) Random scheduler.

(b) Prioritized random scheduler.

Fig. 2. SIRD model results with no social mimicry and a homogeneous population. Panels are divided by scheduler. Colors: infected – green; recovered – red; deceased – blue. Colors apply for all subsequent plots.

Next, we expected a larger deviation from the usual SIRD model if a prioritized random scheduler is used when $\beta = 0.9$ (Fig. 4). For $K = 3$ (Fig. 4a) the differences between the standard SIRD model using either the random or the prioritized random scheduler appear small when agents move one step at a time. However, the differences became apparent when agents could take multiple steps: the effect of the prioritized random scheduler based on agent centrality measures seemed to be an accelerated change of concavity at the expense of not being able to reach extrema as lower as when using the random scheduler. In the case of $K = 9$ (Fig. 4b) the same phenomenon occurred, with the prioritized random scheduler, but not with the random scheduler. Note that the single-step case using the random scheduler is remarkably similar to the multi-step case with the prioritized random scheduler; therefore, we hypothesize that the interaction between agents capable of taking multiple steps and having large network degrees, that is, the interaction between spatial and social areas may be governed by mutual constructive and destructive interference.

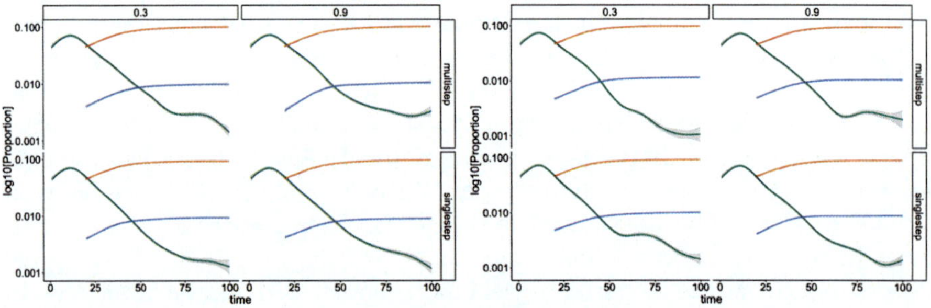

(a) $K = 3$.   (b) $K = 9$.

Fig. 3. Results with social mimicry and a homogeneous population using the prioritized random scheduler, $N = 1000$ agents. Panels are divided by network degree.

Finally, we studied the effect of homogeneous vs heterogeneous motion in agents. Heterogeneous agents sit at extremes of movement across the grid (i.e., different spatial frequency components) as described by Table 2 on page 5. We focused on perturbing the outcomes of Fig. 4b by turning on agent heterogeneity (Figure 5). Heterogeneity accelerates the action in the agent-based model, effectively compressing the timescale at which events occur at the expense of increasing the lower bound of infected. This is not uncommon in stiff dynamical systems in which low and high frequency components interact (Fedorenko 2017). Curiously, the model with heterogeneous agents using the prioritized random scheduler where agents take single steps exhausts the dynamical space faster than its counterparts, and with smaller variability (Fig. 5b). This is most likely a sign of destructive interference between the scheduler choice and the movement behavior of agents. To better understand the effects of choosing a network with a smaller K and smaller β at the opposite extreme, we repeated the analysis with $K = 3$ and $\beta = 0.3$ (Fig. 6). Compression of dynamics still occurs (although not a strongly marked difference), suggesting that it may reach an asymptotic limit when $\beta \rightarrow 1.0$ with little modulation from K. Results suggest that heterogeneity under constant K introduces oscillatory behavior expected due to the system being spatially stiff under the influence of the prioritized random scheduler, while β modulates the dynamics by acting analogously to a dampening force with eventual decay (Fig. 7).

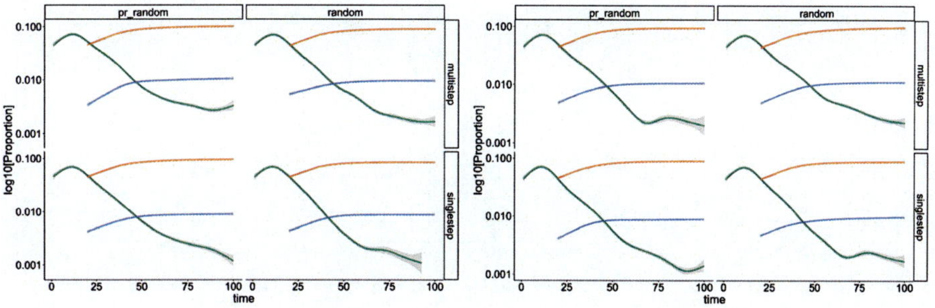

(a) $K = 3$.

(b) $K = 9$.

Fig. 4. Results with social mimicry and a homogeneous population, $N = 1000$ agents, $\beta = 0.9$. Panels are divided by network degree.

We applied the C5.0 classification algorithm (lin Pang and zhang Gong 2009) to understand the relative importance of the model parameters using the data in the Supplementary Table. Here, we set DeathRate = 0.1, corresponding to an SIRD model (64 cases, 20 replicates). Given that the construction of the table was performed by visual inspection in terms of identifying differences and that the number of factors is large in relation to the number of cases, the resulting analysis is far from conclusive (3/64 false positives, 7/64 false negatives, 15.6% total error). However, even this first-pass analysis, the resulting classification tree is informative: mobility homogeneity appears to be the most relevant factor to classify the data (100%), followed by single/multisteps (57.81%), population size (50%), network degree (28.12%), beta (25%) and finally social mimicry (15.62%).

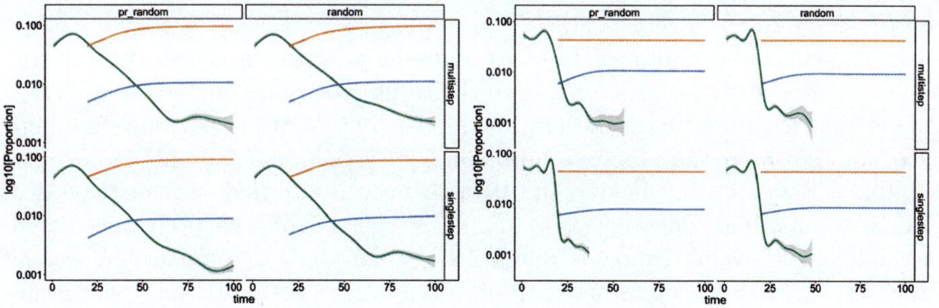

(a) Homogeneous.

(b) Heterogeneous.

Fig. 5. Results with social mimicry, $N = 1000$ agents, $\beta = 0.9$, $K = 9$. Panels are divided by agent heterogeneity.

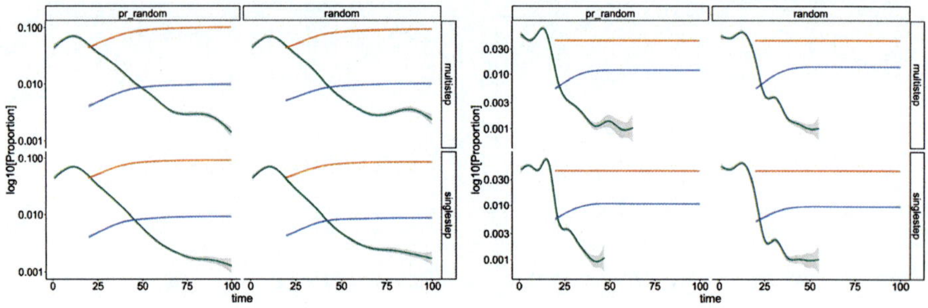

(a) Homogeneous. (b) Heterogeneous.

Fig. 6. Results with social mimicry, $N = 1000$ agents, $\beta = 0.3$, $K = 3$. Panels are divided by agent heterogeneity.

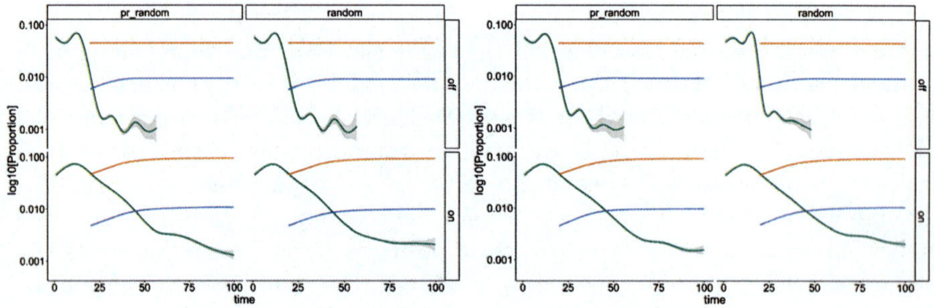

(a) $\beta = 0.3$. (b) $\beta = 0.9$.

Fig. 7. Results with social mimicry, $N = 1000$ agents, $K = 9$. Panels are divided by β.

Bounded by the limitations of this analysis, one may imagine at least two non-competing explanations leading to the result described immediately above. One is that the current results are not representative of any significant difference, and that more sampling resolution on certain factors (e.g., number of steps, population size, network degree and beta) is needed alongside a distance-based comparison method (ongoing work). This is suggestive of a partially automated methodology to evaluate models, one which we plan to develop further. A second explanation, which is consistent with both the latter possibility and the classification tree in Fig. 8, is that of interpreting the frequency of usage of a factor in the classification as inversely proportional to the degree of indirection at which it operates. In this case, social mimicry being the least frequently used becomes the most indirect factor, which, when present, can modulate the action of the other factors synergistically.

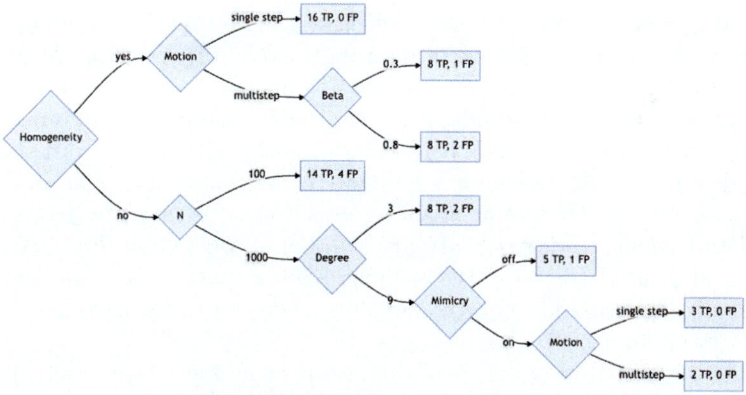

Fig. 8. Classification of simulation scenarios generated via the experimental design described in Table 2 on page 5 using C 5.0. Results are given for true positives (TP) and false positives (FP).

4 Conclusion

In summary, our results show that changing the scheduler in an ABM changes the dynamics significantly depending on how strong the coupling between the properties of the agent and the priority of execution of their actions is. While the conclusions obtained here need additional computational experiments and analysis, the patterns observed so far appear to be consistent with past studies. Choosing between two different schedulers in our model produces deviations with respect to what is expected from a standard contagion model proportional to the magnitude of the infectious peak within an order of magnitude. Our results appear to coincide with the findings of the prior study of fluctuation power spectra in epidemic models on networks, in which resonant fluctuations in stochastic disease models often explain the pattern of incidence for recurrent epidemics (Simoes et al. 2008).

The strength of the coupling, a model-dependent property, appears to be modulated by how a scheduler connects factors across different contexts of action: (1) the spatial modes: how the contagion spreads depending on the range of motion of the agents; and (2) the social modes: how the structure of the network drives non-local consensus about based on local conditions, updating their strategy of motion. Furthermore, each factor appears to play separate, non-orthogonal roles when social mimicry is activated. Having agents move across the grid using multiple steps seems to accelerate the dynamics of the system at the expense of a higher average background of infected agents in the long term. The network degree seems to have a similar effect, but when both operate in tandem the dynamics appears to be driven by some form of constructive interference, suggesting some degree of synchronization between spatial and social contexts. The value of β in general appears to modulate oscillations resulting from heterogeneous motion across the grid.

Several avenues for future work immediately arise from this analysis. First, the analysis of outcomes needs to move from the qualitative to the quantitative via evaluation of non-parametric statistical models operating over computed measures of distance between individual outcomes. Second, a denser sampling of modeling parameters needs to be performed. The above results suggest sampling β more densely within the interval $[0, 0.3]$ to determine how compression of dynamics occurs in conjunction with K, which also needs to be more widely varied. Third, replacing degree centrality by other network centrality metrics can help generalize the character of the coupling between social and spatial contexts in this model. Fourth and final, the model needs to be reproduced across other ABM platforms to show its consistency.

Despite the preliminary nature of this study, two points stand out: (1) scheduler differences matter even for the simplest of ABM models; (2) parameters and interaction protocols of a model interact to suppress or amplify scheduler-dependent model behavior. Admittedly, the descriptive nature of the results reported is a limitation of this study. Nevertheless, the use of C5.0 to classify experimental factors of observed scheduler differences can be used to predict differences in unseen synthetic datasets with different and additional parameter settings. To tease out further the interaction of asynchronous transitions and model parameters, a more systematic inferential approach is necessary, especially to study this for not just this model, or all ABM contagion models, but for general ABM. More generally, the prioritized random scheduler used in the case study is just an implementation constructed to illuminate the interplay between forms of asynchrony and model parameters, i.e., interactions. It was chosen to investigate the ease with which scheduler dependencies in simulations could be generated. Our current choice of schedulers does not exhaust all possible forms of asynchrony in contagion models, let alone in general ABM. Preliminary work suggests that a taxonomy of schedulers is possible and this is currently under investigation.

The point emphasized in this manuscript is not that modeling platforms are incapable of running nontrivial schedules, but that the default schedulers coarse-grain finer details of a model, thereby leading to model misspecification. Furthermore, as past literature has demonstrated, and the current study does too, when and where scheduler dependency arises is not obvious *a priori*. One concrete suggestion of this manuscript and others like it, is to recommend that modelers incorporate multiple schedulers into a typical simulation workflow to determine whether their models' results are significantly sensitive to differences in the activation schedules of their models' agents.

References

Aldous, D.: Interacting particle systems as stochastic social dynamics. Bernoulli **19**(4), 1122–1149 (2013)

Ashton-James, C., van Baaren, R.B., Chartrand, T.L., Decety, J., Karremans, J.: Mimicry and me: the impact of mimicry on self-construal. Soc. Cogn. **25**(4), 518–535 (2007)

Bezanson, J., Edelman, A., Karpinski, S., Shah, V.B.: Julia: a fresh approach to numerical computing. SIAM Rev. **59**(1), 65–98 (2017)
Bonnell, T.R., Chapman, C.A., Sengupta, R.: Interaction between scale and scheduling choices in simulations of spatial agents. Int. J. Geogr. Inf. Sci. **30**(10), 2075–2088 (2016)
Bouré, O., Fatès, N.A., Chevrier, V.: Probing robustness of cellular automata through variations of asynchronous updating. Nat. Comput. **11**(4), 553–564 (2012)
Bromberger, S., Fairbanks, J., et al.: JuliaGraphs/LightGraphs.jl: LightGraphs (2017)
Caron-Lormier, G., Humphry, R.W., Bohan, D.A., Hawes, C., Thorbek, P.: Asynchronous and synchronous updating in individual-based models. Ecol. Model. **212**(3), 522–527 (2008)
Comer, K.W.: Who goes first? An examination of the impact of activation on outcome behavior in agent-based models. Ph.D. thesis, George Mason University (2014)
Cooley, P., Solano, E.: Agent-based model (ABM) validation considerations. In: Proceedings of the Third International Conference on Advances in System Simulation (SIMUL 2011), pp. 134–139 (2011)
Cornforth, D., Green, D.G., Newth, D.: Ordered asynchronous processes in multi-agent systems. Phys. D **204**(1), 70–82 (2005)
Epstein, J.M.: Agent-based computational models and generative social science. Complexity **4**(5), 41–60 (1999)
Fatès, N.: A guided tour of asynchronous cellular automata. In: Kari, J., Kutrib, M., Malcher, A. (eds.) AUTOMATA 2013. LNCS, vol. 8155, pp. 15–30. Springer, Heidelberg (2013). https://doi.org/10.1007/978-3-642-40867-0_2
Fatès, N., Chevrier, V.: How important are updating schemes in multi-agent systems? An illustration on a multi-turmite model. In: Proceedings of the 9th International Conference on Autonomous Agents and Multiagent Systems, vol. 1, pp. 533–540 (2010)
Fedorenko, R.: Stiff systems of ordinary differential equations. In: Numerical Methods and Applications, pp. 117–154. CRC Press (2017)
Frey, D., Šešelja, D.: What is the epistemic function of highly idealized agent-based models of scientific inquiry? Philos. Soc. Sci. **48**(4), 407–433 (2018)
Gallotti, R., Valle, F., Castaldo, N., Sacco, P., De Domenico, M.: Assessing the risks of 'infodemics' in response to COVID-19 epidemics. Nat. Hum. Behav. **4**(12), 1285–1293 (2020)
Hedström, P., Bearman, P.S., Bearman, P.: The Oxford Handbook of Analytical Sociology. Oxford University Press, Oxford (2009)
Kazil, J., Masad, D., Crooks, A.: Utilizing Python for agent-based modeling: the mesa framework. In: Thomson, R., Bisgin, H., Dancy, C., Hyder, A., Hussain, M. (eds.) SBP-BRiMS 2020. LNCS, vol. 12268, pp. 308–317. Springer, Cham (2020). https://doi.org/10.1007/978-3-030-61255-9_30
Klügl, F.: A validation methodology for agent-based simulations. In: Proceedings of the 2008 ACM Symposium on Applied Computing, SAC 2008, pp. 39-43. Association for Computing Machinery, New York (2008)
Lake, M.W.: Explaining the past with ABM: on modelling philosophy. In: Wurzer, G., Kowarik, K., Reschreiter, H. (eds.) Agent-based Modeling and Simulation in Archaeology. AGIS, pp. 3–35. Springer, Cham (2015). https://doi.org/10.1007/978-3-319-00008-4_1
Pang, S., Gong, J.: C5.0 classification algorithm and application on individual credit evaluation of banks. Syst. Eng. - Theory Pract. **29**(12), 94–104 (2009)
Marinoff, L.: The geometry of defection: cascading mimicry and contract-resistant structures. Soc. Philos. Today **17**, 69–90 (2001)

Mistur, E., Givens, J.W., Matisoff, D.: Contagious COVID-19 policies: policy diffusion during times of crisis (2020)

Núñez Corrales, S., Friesen, M., Srikanth, M., Venkatachalapathy, R., Graham, J.: In-Silico models with greater fidelity to social processes: towards ABM platforms with realistic concurrency. In: Yang, Z., von Briesen, E. (eds.) Proceedings of the 2020 Conference of the Computational Social Science Society of the Americas, pp. 155–169. Springer, Cham (2020). https://doi.org/10.1007/978-3-030-83418-0_10

Núñez-Corrales, S., Jakobsson, E.: A generalized theory of interactions for complex multiscale stochastic systems with thermodynamic irreversibility. Bull. Am. Phys. Soc. **65** (2020)

Page, S.E.: On incentives and updating in agent based models. Comput. Econ. **10**(1), 67–87 (1997)

Reia, S.M., Kinouchi, O.: Nonsynchronous updating in the multiverse of cellular automata. Phys. Rev. E **91**, 042110 (2015)

Shu, P., Wang, W., Tang, M., Zhao, P., Zhang, Y.-C.: Recovery rate affects the effective epidemic threshold with synchronous updating. Chaos Interdisc. J. Nonlinear Sci. **26**(6), 063108 (2016)

Simoes, M., da Gama, M.T., Nunes, A.: Stochastic fluctuations in epidemics on networks. J. Roy. Soc. Interface **5**(22), 555–566 (2008)

Watts, D.J., Strogatz, S.H.: Collective dynamics of 'small-world' networks. Nature **393**(6684), 440–442 (1998)

Weimer, C., Miller, J.O., Hill, R., Hodson, D., et al.: Agent scheduling in opinion dynamics: a taxonomy and comparison using generalized models. J. Artif. Soc. Soc. Simul. **22**(4), 1–5 (2019)

Winsberg, E.: Computer simulations in science. In: Zalta, E.N. (ed.) The Stanford Encyclopedia of Philosophy. Metaphysics Research Lab, Stanford University (2019). Winter 2019 edition

Exploring the Impact of Social Network Density and Agent Openness on Societal Polarization

Justin Mittereder, Robert S. W. Carroll, Brandon Frulla, and Stephen Davies(✉)

Department of Computer Science, University of Mary Washington, 1301 College Avenue, Fredericksburg, VA 22401, USA
{jmittere,rcarrol2,bfrulla,stephen}@umw.edu

Abstract. We present an agent-based model, inspired by the opinion dynamics (OD) literature, to explore the underlying behaviors that may induce societal polarization. Our agents interact on a social network, in which adjacent nodes can influence each other, and each agent holds an array of continuous opinion values (on a 0–1 scale) on a number of separate issues. We use two measures as a proxy for the virtual society's "polarization:" the average assortativity of the graph with respect to the agents' opinions, and the number of issues on which agents have persistent disagreement even after the model reaches an equilibrium.

We look at two model parameters that affect polarization. The first is the density of edges in the network: this corresponds to the average number of meaningful social connections that agents in a society have. Contrary to our early hypothesis, we find that lower edge density results in higher levels of assortativity. The second is the "openness" of agents to differing opinions; *i.e.*, how close a neighboring node's opinion on an issue must be to an agent's own before the agent will adjust its opinion on a different issue. We refer to this novel mechanism as cross-issue influence. Through this mechanism, we find that when agents in the model are less open to new opinions, there will be less consensus on any given issue for all agents in the model.

CCS Concepts
• **Computing methodologies → Modeling and simulation; Agent/discrete models.**
ACM Reference Format:
Justin Mittereder, Robert S. W. Carroll, Brandon Frulla, Stephen Davies. 2021. Exploring the Impact of Social Network Density and Agent Openness on Societal Polarization. In Proceedings of CSSSA's Annual Conference on Computational Social Science (CSS'21). ACM, New York, NY, USA, Article 4, 7 pages. https://doi.org/10.475/123_4.

Z. Yang and E. von Briesen (Eds.): CSSSA 2021, SPCOM, pp. 71–84, 2022.
https://doi.org/10.1007/978-3-030-96188-6_6

Keywords: Opinion dynamics · Echo-chambers · Binary voter model · Social networks · Polarization

1 Introduction

The recent events that transpired at the U.S. Capitol on January 6th were a vivid reminder of the deep divide within the nation. There are signs that the United States is experiencing political polarization now like it has never seen before. As individuals stormed the Capitol, Americans watched in horror. Although this singular event is now in the past, the underlying tension that preceded it still remains.

Polarization – reflected in echo chambers, entrenched views, and the vilification of those whose opinion differs – can be harmful to a democratic society. It can inhibit the reaching of consensus and compromise upon which a democracy is built, and can result in even greater amounts of damage than what ensued in the U.S. on January 6th if left unchecked. Further, polarization affects not only political actors, but also the interpersonal relationships among the rank and file citizens of a country which bolster and strengthen society.

In this paper, we look at two societal variables that we believe may significantly impact polarization in a society. The first is the density of social connections: in other words, the average number of social ties a member of that society has. The second is the degree of "openness" in the society: namely, how willing its members are to consider changing their views. We suspect that both of these factors play a role in determining the aggregate polarization of a society.

In order to explore these phenomena, we created an Agent-Based Model (ABM) of heterogeneous agents in the spirit of much of the Opinion Dynamics (OD) literature. These agents interact with each other on a random, static social network and change their opinions on issues over time based on the opinions of their network neighbors. One novel feature of our model, termed cross-issue influence, is the way agents influence one another: one agent will not allow another agent to influence its opinion on an issue unless the two agents already have sufficient agreement on another (randomly chosen) issue. The justification for this is related to the well-known observation of "homophily" in social psychology: people are prone to trust those who already agree with them on something, and hence are more likely to be persuaded by them on other matters.

The goal of our research is to determine what micro behaviors of individuals are sufficient to produce a change in the degree of political polarization in the society. As explained below, we choose to measure polarization in two different ways: the average similarity of an agent to its neighbors (called "assortativity" in social network terminology), and the likelihood that no consensus will be reached on an issue (called opinion "clustering" in the OD literature).

2 Related Work

2.1 Opinion Dynamics

Opinion Dynamics models seek to reproduce the phenomenon of individual agents forming opinions over time via mutual influence. They allow the researcher to explore the macro-level patterns that may arise in a society from a set of simple influence rules defined on the micro-level. For instance, the Binary Voter Model (BVM), the original and perhaps most influential OD model [3,16], features a set of interacting agents, each of which holds a binary opinion. The single influence rule is that agents periodically change their opinion to match one of their influencers, chosen at random. Among other things, the model demonstrates that such a system will always eventually reach uniformity of opinion.

Many researchers (*e.g.*, [13,24]) have expanded this idea to model continuous, rather than discrete, opinions: these are typically expressed as real numbers between 0 and 1. In addition to better capturing the nuance of real-life viewpoints (which are not usually completely black or white on any issue), continuous opinions lead naturally to incorporating a form of **homophily** [19] into the model: agents will only choose to be influenced by agents whose existing opinion is already close to their own. Termed "**bounded confidence**" (BC) by [15], this feature can result in non-convergence to uniformity depending on the value of the threshold agents use to gate influence [7,15,23]. The term "**clustering**" (or "opinion clustering") has been used to describe the resulting equilibrium reached by such models, in which subsets of the agents each converge on a different opinion value and are henceforth no longer persuadable by other agents.

A smaller number of studies have considered "**multidimensional opinions**," in which each agent maintains a separate opinion on each of several different "issues" rather than on just one.[1] The opinions in a multidimensional setting have been modeled as discrete [7] or even as boolean variables combined in arbitrary logic formulas [1,2]. Oddly, modeling multidimensional opinions as an array of continuous values is rarely seen. One purpose of using multidimensional opinions could be to see how an agent's opinions on different issues interact with one another. This is explored by the boolean expressions in [1] and [2]; in [7] the multidimensional opinion for each agent is instead used merely as an element in a vector space whose (Hamming) distance from other agents' multidimensional opinions can be computed and compared to a BC threshold.

With respect to these previous efforts, our model resembles the BVM but gives the agents continuous, multidimensional opinions. Our model also implements BC, but in a different way than models like [23] do: before accepting the

[1] This is to be carefully distinguished from "**opinion vectors**," which represent an agent's degree of support for each of several alternatives on the *same* issue. (See, *e.g.*, [21].) Unlike multidimensional opinions, these opinion vectors are often restricted to be members of a probability simplex.
To be concrete about the difference, an agent in a model with multidimensional opinions might have a value of .8 for the "pro-gun control" issue, .9 for the "raise the minimum wage" issue, and .4 for the "restrict fracking" issue. By contrast, an agent in a model with opinion vectors might have a value of .2 for the "raise taxes to fund infrastructure" alternative, .7 for the "cut military spending to fund infrastructure" alternative, and .1 for the "increase IRS audits to fund infrastructure" alternative, all possible solutions to the single "how to fund infrastructure" issue. In the latter case, the options are considered mutually exclusive and must sum to 1 for any agent. (Of course, the specific real-world examples here are only for illustration; OD models represent "issues" and the "opinions" about them completely abstractly.).

influence of a fellow agent on an issue, an agent in our model must already be close in opinion to that agent on a *different* issue. This mechanism we refer to as cross-issue influence is meant to mimic a phenomenon of human behavior: if I learn that your viewpoint on issue A is close to my own, homophily suggests that I will trust you, and I will therefore be willing to consider your viewpoint on issue B. To our knowledge, this mechanism of agent influence has not been previously explored.

2.2 Polarization

Polarization can mean different things to different people; we therefore begin by briefly establishing a dictionary of terms that we will refer back to throughout this paper.

Arguably the most familiar manifestation of polarization – which we term "**diametricity**" – is when a group experiences opinions shifting away from common ground to polar sides, leaving nobody 'in the middle' on a specific issue. We do not study this flavor of polarization in this paper.

We use the graph theory term **assortativity** to represent a second type of polarization, which is rooted in the tendency that people have to form connections with people who have similar views. This idea is supported by [18] which focuses on physical proximity breeding connections, as well as [2] which states that we are more likely to form connections with those that we already are in agreement with on another issue. The well-known concept of homophily comes into play here, as studied in [5] and [22]. Assortativity is a way to quantify the presence of "echo chambers" in a society, in which people are exposed mostly (or solely) to opinions that confirm what they already believe [4,10].

Finally, a third form of polarization is one that can be measured as follows: how often do opinions on issues result in **clustering**? For example, if all individuals had the same belief, there would be one opinion cluster. However, in a polarized society, there are clusters of opinions for any given issue. In this way, higher clustering in a society represents when individuals are entrenched and no longer willing to change their opinion on a given issue. The mechanism that we use in this paper to calculate the number of opinion clusters will be explained later.

Regarding the role of a society's "openness," one question that arises is the psychological basis for this attribute. Which personality trait plays the biggest role in an individual's likelihood to change their opinions on a particular issue? The 'Big 5' personality trait group [17], well-researched since the 1980s, contains Openness-to-Experience (OE) as one of its five traits. OE can be defined as "cognitive flexibility" [8], or "[openness can be] associated with having a vivid imagination and [...] receptivity to one's own and other's emotions; a willingness to try new experiences" [12]. As the research shows, openness plays a crucial role in an individual's ability to relate to others, as well as to consider adopting outside ideas as their own. More than the Big 5's Agreeableness and Conscientiousness traits, OE seems to reflect well what our model is attempting to capture.

3 Variables

In this section we define the two important independent variables whose effect on the model's behavior we seek to discover, and the two dependent variables we measure at simulation's end.

3.1 Independent Variables

3.1.1 Openness As mentioned earlier, research shows that **openness** plays a crucial role in an individual's ability to relate to others, as well how easily they adopt outside ideas as their own. To quantify this as a model parameter, we incorporate openness as a threshold on a continuum from 0 to 1; this threshold is used to compare agent opinions during their pairwise interactions. Low levels of openness produce models in which agents only very rarely change their opinions (namely, only when encountering neighboring agents whose opinion on another issue is very close to their own). High levels produce models in which agents eagerly incorporate the opinions of others on almost every interaction.

3.1.2 Edge Probability The other parameter represented in our model is the *density* of social connections. To implement the concept of different degrees of social connection, we used the Erdös-Rényi graph generation algorithm to generate a random graph of connected nodes. With the Erdös-Rényi graph generation algorithm, we can specify the **edge probability** which represents the probability that there will be an edge between any two given nodes. Using the edge probability, we can control the density of the resulting graph. As a result, edge probability directly corresponds to the density of social connections in our model.

3.2 Dependent Variables

3.2.1 Graph Assortativity One way we measure the simulated society's polarization is through the resulting network's "assortative mixing," or simply graph **assortativity**. This represents the degree to which an agent's opinions will have similar values to those of its network neighbors, on average.

The assortativity of a network has a value between -1 and 1, where 1 indicates "perfect assortative mixing" – *i.e.*, a situation where every agent's opinions are identical to each of its graph neighbors'. An assortativity of 0 indicates that the agents' social connections have no correlation at all with their opinion values: having a social tie with another agent does not mean an agent is any more (or less) likely to have opinions similar to that agent. This will be approximately true when the model is initialized and before the iterative process begins. (Negative assortativity values correspond to networks in which an agent is *less* likely to agree with its network neighbors than with agents in general.)

Assortativity is thus a way to measure the extent to which agents become surrounded by (only) like-minded agents, and are therefore no longer exposed to alternative points of view. Since we need to obtain the graph's assortativity with respect to *multiple* attributes (*i.e.*, the opinions an agent has on all of the issues), we simply compute the network's assortativity for each issue separately (as defined in [20], p.5) and average it over all the issues.

3.2.2 Opinion Clustering The second dependent variable of our model is opinion **clustering**. This measures how often the opinions that agents have on a given issue fail to converge to a uniform value, instead remaining bifurcated among two or more values in perpetuity. Each group of agents who, at simulation's end, have the same opinion on an issue (within some small tolerance ϵ) are termed an "opinion cluster" (a term used by [11]) on that issue.

For clarity, we refer to any issue on which all agent opinions eventually converge to the same value as a "**uniform issue**," and any issue that instead produces opinion clusters as a "**clustered issue**."

One challenge is defining what qualifies as an clustered issue, given that agent opinions are represented as real numbers that may asymptotically converge to, but never actually reach, the same value. We use the following mechanism. To calculate the number of clusters for an issue, we add agents to a cluster after every step of the model. If the absolute value of the difference between an agent's opinion and the average opinion of a pre-existing cluster is within a threshold (0.05), the agent is added to that cluster. If this is not the case, the agent is added to a new cluster in which it is the first occupant.

4 Model

The model is presented using an abbreviated version of the ODD protocol [14].

4.1 Purpose

The model simulates interactions on a random social network of agents, each with an array of continuous, numeric **opinion** attributes. Its purpose is to investigate the way in which two factors contribute to the emergence of polarization in the network: the **edge_probability**, a value reflecting the density of social connections in the network; and the **openness**, a value representing how closely one of an agent's opinions must be to that of a potential influencer in order to accept influence. (See Sect. 4.3, below.)

Using the model, we hope to gain general insight on the emergence of this polarization within social networks and how different parameters affect this.

4.2 Entities, State Variables and Scales

The entities within the model are Agents, having the following attributes:

ID. A unique ID for the agent.

Opinions. An array of numbers, representing opinions on issues, each having a value between 0 and 1. This represents the degree to which the agent "agrees" or "disagrees" with an issue, with 0.5 being neutral.

Neighbors. A subset of the other agents in the model, to whom this Agent has a social connection. The entire set of Agents and their social connections form an undirected graph (*i.e.*, all social connections are bidirectional) and the graph is fixed throughout the simulation.

4.3 Process Overview and Scheduling

After the model has been initialized, the following sequence is executed for each of a fixed number of steps in the simulation:

(1) An agent X is chosen at random.
(2) A neighbor of X (call it Y) is chosen at random.
(3) An issue I_1 is chosen at random.
(4) The absolute difference between X's opinion on I_1 and Y's opinion on I_1 is measured.
(5) Another opinion I_2 ($\neq I_1$) is chosen at random.
(6) If the difference between X's and Y's opinion on issue I_1 is less than or equal to the model's **openness** parameter, set X's opinion on I_2 to be the average of X's and Y's current I_2 opinions.

4.4 Initialization

The simulation is initialized with 50 agents, each having 5 opinions set to independent uniform random values between 0 and 1. The agents are then connected to each other using a random undirected Erdös-Rényi graph [9] with parameters $N = 50, p =$ **edge_probability**. (If the graph is not connected, a new random graph is generated until a connected one is obtained.)

The model's step limit is usually set to 150, as most change in the agent's opinions after 150 steps is negligible.

5 Hypothesis

We form the following hypotheses about the model's behavior.

Hypothesis 1a: (H_{1a}). Mean assortativity will increase with the edge probability of an Erdös-Rényi graph.
Hypothesis 1b: (H_{1b}). Mean assortativity will increase when the openness parameter of agents in the model is lower.
Hypothesis 2a: (H_{2a}). The number of clustered issues will be negatively correlated with the edge probability of an Erdös-Rényi graph.
Hypothesis 2b: (H_{2b}). The number of clustered issues will increase when the openness parameter is lower for all agents in the model.

For H_{1a}, we hypothesize that increasing the connectivity of an Erdös-Rényi graph by raising the edge probability will result in higher assortativity. This hypothesis is based mainly on real-world observations: the number of social connections available to those with Internet access has increased in the past few decades (due to social media [6]), and the degree of homophily exhibited in members of a social circle has also (at least anecdotally) increased. Since both the density of connections and the homophily of those joined by such connections has increased in the real world, we presume the same effect will follow in our model.

For H_{1b}, we hypothesize that when agents in the model are less open to new opinions, there will be a higher average assortativity and therefore polarization. When all agents have a lower level of openness, they will only be interacting with agents that have opinions similar to their own; therefore, we expect to see higher levels of assortativity.

For H_{2a}, we assume that raising the connectivity of an Erdös-Rényi graph by increasing the edge probability will result in fewer clustered issues. As a graph becomes more densely connected, agents will have a wider variety of neighbors to receive influence from. As a result, agents should merge to the consensus opinion for any given issue more often in a more densely connected graph.

For H_{2b}, we believe that lowering the openness parameter of agents in the model will result in more clustered issues across the model. When agents are less open to distant opinions, there will be more variety of opinion for any given issue.

6 Results

6.1 H_{1a} and H_{1b}

To test H_{1a}, we first establish a model with 50 agents, 5 issues, and an openness parameter of 0.40. In order to measure the impact of varying edge probability on average assortativity across all issues, we run each combination of parameters 20 times starting with an edge probability of 0.05 and ending with an edge probability of 0.95, in increments of .05. The results of this model run are shown in Fig. 1.

We see that as the edge probability (or density of social connections) increases, the average assortativity across all issues decreases. This is the exact opposite of our hypothesis. One possible explanation for this result is that when connections are more dense, there is a higher chance that agents will be exposed to a more diverse set of opinions. There is thus a higher chance that agents will be pulled to the 'average' opinion for a given issue, which would produce lower assortativity. From this finding, we may be able to infer that societies where individuals are more densely connected may experience less polarization than more sparsely-connected societies do.

In addition to the negative correlation between density of social connections and polarization, we also see that the relationship between these two variables appears negative-exponential in nature. The variance was too high, however, for

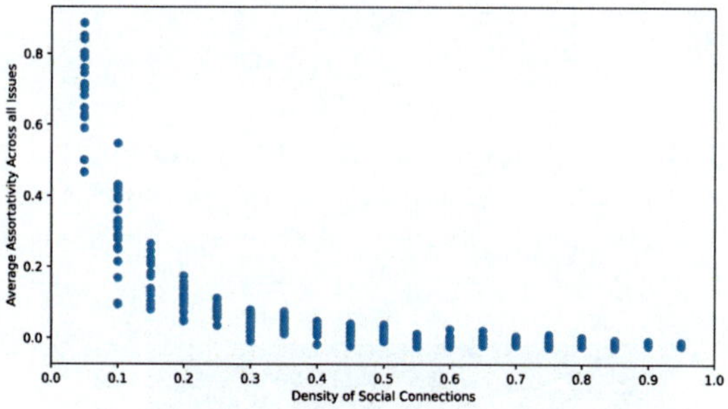

Fig. 1. Average assortativity across all issues and edge probability

us to draw a solid conclusion on whether the relationship truly conforms to a negative-exponential, a power-law, or any other standard distribution.

To test H_{1b}, we first establish a model with 50 agents, 5 issues, and an edge probability of 0.50. In order to measure the impact of varying the openness parameter on average assortativity across all issues, we ran each combination of inputs 20 times with an openness parameter ranging from 0.05 to 0.95 in increments of .05. The results of this model run are shown in Fig. 2. As is depicted, there is no obvious relationship at all between the openness parameter and the average assortativity across all issues.

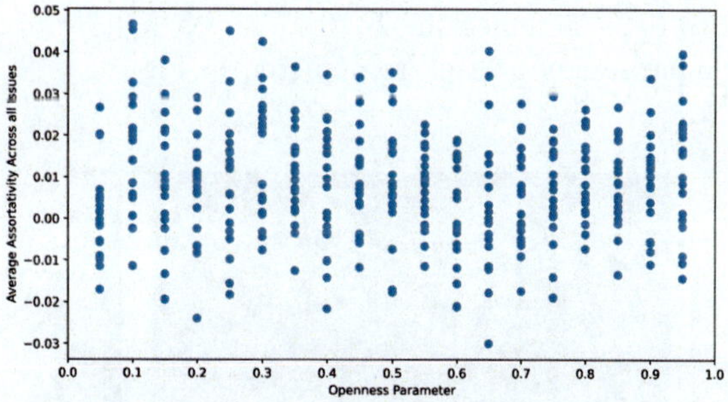

Fig. 2. Average assortativity across all issues and openness parameter

This is an interesting result. Agents in the model are influenced when they are close in opinion (within our openness parameter) to another agent on the same issue. Therefore, we believed that openness would play a role in determining the assortativity of a society. It should be noted that we tested this hypothesis with multiple different values of the edge probability (0.15, 0.40, and 0.50), to ensure that the edge probability was not having an impact on our results. Even still, we hope to investigate this hypothesis further in future research.

6.2 H_{2a} and H_{2b}

To test H_{2a}, we establish a model with 50 agents, 5 issues, and an openness parameter of 0.30. First, we ran a parameter sweep varying the edge probability from 0.05 to 0.95 to measure the impact of this parameter on the number of opinion clusters. The results of this parameter sweep are shown in Fig. 3.

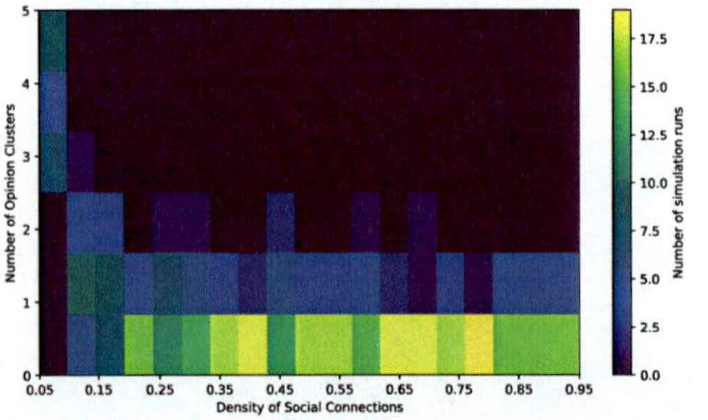

Fig. 3. Number of opinion clusters and edge probability (0.05–0.95)

We noticed that as with H_{2a}, there appears to be a tipping point with the number of opinion clusters and the edge probability. To further explore this hypothesis, we ran another parameter sweep, this time varying the edge probability from 0.05 to 0.40 incrementing by 0.01 for each suite of 20 model runs. The results of this parameter sweep are depicted in Fig. 4.

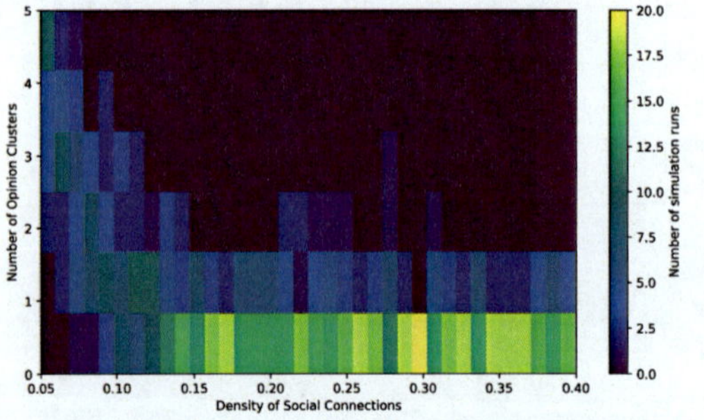

Fig. 4. Number of opinion clusters and edge probability (0.05–0.40)

Our results confirm H_{2a}; the number of opinion clusters and edge probability have a negative relationship. We believe this may be explained by the implications of a high density for a graph of nodes. For example, when a graph of 50 nodes has a density of 0.05, the average number of social connections will be 2.5. We are able to calculate the average number of social connections by multiplying the chance there will be an edge between any two nodes (edge probability) and the number of nodes. When the edge probability, or density of the graph, increases slightly to 0.2, the average number of social connections will rise to 10 connections. As a result, the geodesic distance between two nodes decreases rapidly because each node is proportionately connected to more nodes in the graph. This may reveal why we saw that only a certain level of density is required for the number of opinion clusters to drop sharply. Undeniably, a tipping point exists with the number of opinon clusters when increasing the density of an Erdös-Rényi graph in our model.

To test H_{2b}, we establish a model with 50 agents, 5 issues, and an edge probability of 0.50. First, we ran a parameter sweep varying the openness parameter from 0.05 to 0.95 to measure the impact of varying the openness parameter on the number of opinion clusters. The results of this parameter sweep are shown in Fig. 5.

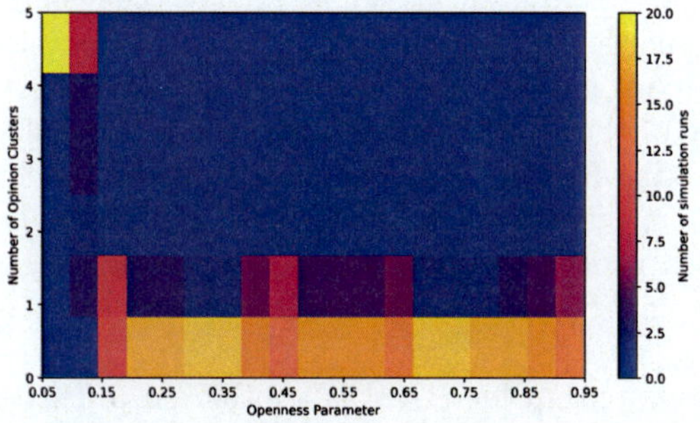

Fig. 5. Number of opinion clusters and the openness parameter (0.05–0.95)

We noticed that there was little to no difference between an openness parameter of 0.5 and 0.7. However, we observed that the openness parameter had more impact on the number of opinion clusters when the parameter was closer to 0.10. To further explore this result, we ran another parameter sweep with 50 agents, 5 issues, an edge probability of 0.50, and a suite size of 20. This time, we varied the openness parameter from 0.05 to 0.40. Our results are depicted in Fig. 6.

This graph indicates that there is a tipping point for the openness parameter. When the openness for agents in the model is very low, the agents did not agree on many issues. However, as Fig. 6 indicates, when we increase the openness parameter slightly, the number of opinion clusters across the model quickly drops. As a result, we can infer that low levels of openness in a society may induce more polarized societies. When agents in the model are less open to distant opinions, there are more opinion clusters for any given issue. However, the tipping point leads us to believe that slightly higher levels of openness are sufficient to reach uniformity on a given issue for all agents in the model. To conclude, when using the cross-issue influence mechanism, marginally higher levels of openness led to less polarization in the model.

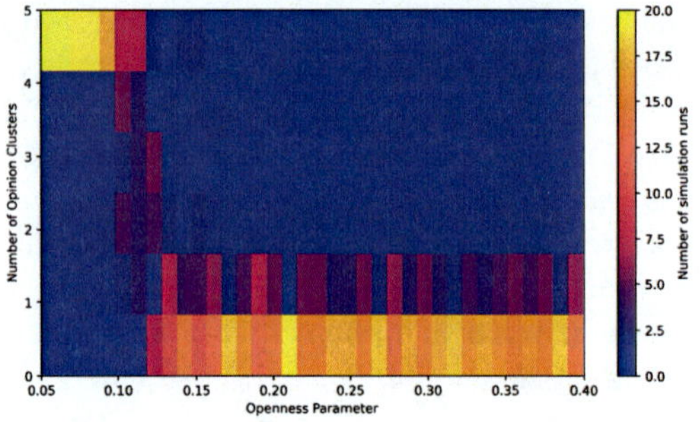

Fig. 6. Number of opinion clusters and openness (0.05–0.40)

7 Discussion and Future Work

Multiple results in this research surprised us. Firstly, the results from testing H_{1a} did not reflect our anecdotal experiences. When increasing the density of a society's connection, we instead saw *lower* assortativity. We believe this may be due to the static nature of the model's social network. In the real world, homophily not only causes existing friends to become more like each other, but also causes people to select (or reject) friends based on their similarity. In future work, we intend to add this feature to the model, producing a dynamic graph, and discover whether this addition is sufficient to produce a positive density/assortativity relationship.

The lack of a relationship for H_{1b} was another surprising result. We extensively tested this hypothesis, but the results did not indicate any statistically significant relationship. This result remains unexplained.

The tipping points observed when testing hypotheses H_{2a} and H_{2b} were compelling results. When even slightly increasing the density of a graph, the number of clustered issues can drop quickly. This would seem to indicate that the degree to which a society forms consensus can be quite sensitive to the average number of social connections people maintain, at least within a certain range. Too, the openness of a society's members – however that might be quantified in a real population – produced an even steeper tipping point. One interpretation would be that even small changes in the tolerance people have for dissenting views can produce great gains in reducing polarization. We also plan to investigate the behavior of models with agents that are heterogeneous with respect to openness, since OE and other traits are obviously not uniform across a real population.

Another mechanism we hope to explore more in future research is cross-issue influence. This concept is an extension of Hegselmann and Krause's bounded-confidence mechanism. In this research we explore cross-issue influence with only attracting forces. However, we hope to investigate the results that would be produced when a repelling force is implemented into the cross-issue influence mechanism. Rather than only having agents move closer to one another on issue X, we could also have them be pushed away from each other on issue X if they disagree above a certain threshold on a separate issue Y.

References

1. Battistella, E., Cholvy, L.: Modelling and simulating extreme opinion diffusion. In: van den Herik, J., Rocha, A.P. (eds.) ICAART 2018. LNCS (LNAI), vol. 11352, pp. 79–104. Springer, Cham (2019). https://doi.org/10.1007/978-3-030-05453-3_5
2. Cholvy, L.: Diffusion of opinion and influence. In: Schockaert, S., Senellart, P. (eds.) SUM 2016. LNCS (LNAI), vol. 9858, pp. 112–125. Springer, Cham (2016). https://doi.org/10.1007/978-3-319-45856-4_8
3. Clifford, P., Sudbury, A.: A model for spatial conflict. Biometrika **60**(3), 581–588 (1973). https://doi.org/10.2307/2335008
4. Dandekar, P., Goel, A., Lee, D.T.: Biased assimilation, homophily, and the dynamics of polarization. Proc. Natl. Acad. Sci. **110**(15), 5791–5796 (2013)
5. Davies, S.: The twin impact of homophily and accessibility on ideological polarization. In: Proceedings of the 2017 International Conference of The Computational Social Science Society of the Americas, pp. 1–8 (2017)
6. Dean, B.: How Many People Use Social Media in 2021? (65+ Statistics) (2021). https://backlinko.com/social-media-users
7. Deffuant, G., Neau, D., Amblard, F., Weisbuch, G.: Mixing beliefs among interacting agents. Adv. Complex Syst. **3**(01n04), 87–98 (2000). http://www.worldscientific.com/doi/abs/10.1142/S0219525900000078
8. DeYoung, C.G., Peterson, J.B., Higgins, D.M.: Sources of openness/intellect: cognitive and neuropsychological correlates of the fifth factor of personality. J. Pers. **73**(4), 825–858 (2005). https://doi.org/10.1111/j.1467-6494.2005.00330.x
9. Erds, P., Rnyi, A.: On random graphs, I. Publicationes Mathematicae (Debrecen) **6**, 290–297 (1959). http://www.citeulike.org/group/3509/article/4012374
10. Flaxman, S., Goel, S., Rao, J.M.: Filter bubbles, echo chambers, and online news consumption. Public Opin. Q. **80**(S1), 298–320 (2016)
11. Fotakis, D., Palyvos-Giannas, D., Skoulakis, S.: Opinion dynamics with local interactions. In: Proceedings of the Twenty-Fifth International Joint Conference on Artificial Intelligence (IJCAI-16), p. 7 (2016)
12. Furnham, A., Cheng, H.: Childhood intelligence predicts adult trait openness: psychological and demographic indicators. J. Individ. Differ. **37**(2), 105–111 (2016). https://doi.org/10.1027/1614-0001/a000194
13. Ghaderi, J., Srikant, R.: Opinion dynamics in social networks: a local interaction game with stubborn agents. arXiv:1208.5076 [cs] (2012)
14. Grimm, V., et al.: A standard protocol for describing individual-based and agent-based models. Ecol. Model. **198**(12), 115–126 (2006). https://doi.org/10.1016/j.ecolmodel.2006.04.023
15. Hegselmann, R., Krause, U.: Opinion dynamics and bounded confidence models, analysis, and simulation. J. Artif. Societies Soc. Simul. **5**(3) (2002). http://citeseerx.ist.psu.edu/viewdoc/download?doi=10.1.1.454.3597&rep=rep1&type=pdf
16. Holley, R.A., Liggett, T.M.: Ergodic theorems for weakly interacting infinite systems and the voter model. Ann. Prob. **3**(4), 643–663 (1975). https://doi.org/10.1214/aop/1176996306
17. John, O.P., Srivastava, S.: The Big-Five Trait Taxonomy: History, Measurement, and Theoretical Perspectives, vol. 2. University of California Berkeley (1999)
18. Klinkner, P.A., Hapanowicz, A.: Red and Blue Dj Vu: measuring political polarization in the 2004 election. Forum: J. Appli. Res. Contemp. Polit. **3**(2) (2005). https://doi.org/10.2202/1540-8884.1079

19. McPherson, M., Smith-Lovin, L., Cook, J.M.: Birds of a feather: homophily in social networks. Ann. Rev. Sociol. 415–444 (2001). http://www.jstor.org/stable/10.2307/2678628
20. Newman, M.E.J.: Mixing patterns in networks. Phys. Rev. E **67**(2), 026126 (2003)
21. Srbu, A., Loreto, V., Servedio, V., Tria, F.: Opinion dynamics with disagreement and modulated information. J. Stat. Phys. **151**(1/2), 218–237 (2013). https://doi.org/10.1007/s10955-013-0724-x
22. Taylor, C.E., Mantzaris, A.V., Garibay, I.: Exploring how homophily and accessibility can facilitate polarization in social networks. Information **9**(12), 325 (2018)
23. Tsang, A., Larson, K.: Opinion dynamics of skeptical agents. In: Proceedings of the 2014 International Conference on Autonomous Agents and Multi-agent Systems. In: International Foundation for Autonomous Agents and Multiagent Systems, pp. 277–284 (2014). http://dl.acm.org/citation.cfm?id=2615778
24. Weisbuch, G., Deffuant, G., Amblard, F., Nadal, J.P.: Interacting Agents and Continuous Opinions Dynamics. arXiv:cond-mat/0111494 (2001)

Learning Actor Preferences by Evolution

H. Van Dyke Parunak(✉)

Parallax Advanced Research, Beavercreek, OH 45431, USA
van.parunak@gmail.com

Abstract. A critical challenge in modeling social behavior is fitting model parameters to behaviors observed in the real world. In some applications, we have been able to evolve appropriate model components against a fitness function that compares simulated behavior against observed behavior. We outline scenarios where this capability is useful, review past work in this area, and describe current and planned experiments in evolving SCAMP models against observations. In addition to learning agent preferences successfully, our experiments demonstrate the possibility of generating realistic synthetic data for experimentation, and give deeper insight into the behavior of the SCAMP system.

Keywords: Agent based modeling · Genetic algorithm · Model fitting

1 Introduction

Social simulations are often used to help policy makers and social scientists understand the real world. Constructing such a model forces researchers to articulate their assumptions, and allows experiments that would be impractical or unethical to perform in real life. But the value of such models depends on how well they reflect the real world.

Sometimes it is appropriate to assess a model's fit with reality after it has been constructed and run. After all, getting the structure and parameters of a model correct is the responsibility of the modeler. Any social model embodies a theory of the social science underlying the scenario being modeled. The modeler embeds that theory in the model during construction, and comparison of the model's output with reality helps evaluate the theory that supported its construction.

But there are aspects of a socially and psychologically realistic model that are difficult to know a priori, such as the preferences held by different actors, or the goals being pursued by the groups with which they identify. If representatives of these groups are available, interview techniques can derive such information, but the process is costly, and may not be applicable if one is modeling participants in a conflict, who may try to obscure their motives for competitive advantage. In such cases, we need mechanisms to tune the model to actual observed behavior. One promising approach to this problem is the use of stochastic iterative optimization. This term [2] includes any optimization process that repeatedly evaluates the fitness of one or more candidate solutions and updates them, typically by randomly changing the best individuals (mutation) or by sharing information among them (in a genetic algorithm, crossover; in particle swarm optimization, attraction

Z. Yang and E. von Briesen (Eds.): CSSSA 2021, SPCOM, pp. 85–97, 2022.
https://doi.org/10.1007/978-3-030-96188-6_7

to nearby individuals). Examples include ant colony optimization, differential evolution, particle swarm optimization, and artificial immune systems, among others. The work reported here uses a genetic algorithm, so we use the term "evolution", but include some work using particle swarms in our discussion.

The success of evolutionary tuning depends on the elements that a modeling technology exposes as potential chromosomes for the tuning algorithm. The SCAMP causal modeling system (Social Causality using Agents with Multiple Perspectives) that we[1] introduced at CSS 2020 offers some simple yet expressive structures that are ideally suited to this purpose.

This paper begins by reviewing some previous applications for which stochastic iterative optimization has been useful (Sect. 2). We then briefly review the structure of SCAMP to highlight features that lend themselves to exploration with genetic methods (Sect. 3), and discuss our work in evolving one particular feature, the agent's preference vector (Sect. 4). Section 5 offers some concluding discussion.

2 Some Use Cases for Behavior Evolution

Evolutionary methods have been described as "the second-best method to solve any problem", the best method being one that takes into account the specific structure of the problem. For example, one could use population search to find the maximum of an analytic equation, but would be better advised to start with setting the first derivative to 0, constraining the second to be negative, and solving. But we often do not have a compact, analytically tractable description of the system, and iterative optimization performs surprisingly well. Here are some examples.

Often, a set of single-valued *parameters* are available. Many systems offer ways to sweep the parameter space of a model systematically (for example, the Repast batch facility; NetLogo's Behavior Space), but as the number of parameters increases, the number of required runs quickly becomes prohibitive, and population-based optimization can home in on usable parameters with far fewer runs than exhaustive sweeps. Both Scikit-Learn (for Python) and Mathematica include evolutionary frameworks that can be wrapped around functions written in those languages. The EvoPER package [3] offers system-level support for stochastic optimization of the parameters of a Repast or NetLogo model, using particle swarm optimization (PSO), simulated annealing (SO), and ant colony optimization (ACO), but not chromosome-inspired methods such as genetic algorithms.

We optimized parameters with evolution in the DARPA JFACC program to plan paths for air mission packages [14]. The physical mission package (a team of aircraft) repeatedly sends out software agents to search a digital map of the terrain that records known targets and threats in the form of pheromones. Four digital pheromones are used: *RTarget* (deposited by an agent representing a target), *RThreat* (deposited by an agent representing a threat), *GTarget* (deposited by an agent that has encountered a target and is returning to the mission package), and *GNest* (deposited by an agent that has left the

[1] The team included J.A. Morell of 4.699 LLC; L. Sappelsa of ANSER LLC; J. Greanya and S. Nadella of Wright State Research Institute. Kathleen Carley of CMU consulted on social networks. J. Greanya produced the evolutionary tools used in this paper.

mission package and is seeking a target). The agent may also know *Dist*, an estimate of the distance to the target when the target location is known. Manual experimentation yields an equation to combine these factors into an attractive force for the aircraft:

$$\frac{\theta \cdot RTarget + \gamma \cdot GTarget + \beta}{(\rho \cdot GNest + \beta)(Dist + \varphi)^{(\delta + \alpha(RThreat+1)))} + \beta}$$

The Greek letters are all tunable parameters. A genetic algorithm greatly outperforms both manual tuning and random search in solving a variety of complex threat-target scenarios, including one that requires finding a route around along array of missile batteries into a long gauntlet to reach the target.

We used particle swarm optimization to tuning a set of eight parameters in searching for phase shifts in a distributed multiagent algorithm for the graph coloring problem [1]. Some of these parameters describe the actions of the agents (probability of a color change, message transfer time, the degree to which change is constrained by the local degree of conflict, and the mechanism used to select the next color), while others describe the graph (the number of nodes in the graph, neighbors per node, and available colors, and the mechanism used to construct the graph).

Contributors to the DARPA RAID program produced adversarial reasoning modules (ARMs) to anticipate the actions of enemy troops in urban combat. The ARMs were tested in human-staffed war games [7]. Our system was a swarming simulation of the battlespace, similar to the geospatial component of SCAMP. In predicting the movement of enemy troops, we used evolution in two ways [8].

First, in any complex system, prediction becomes unreliable beyond a short horizon, as the system's dynamics lead to divergence of trajectories [10]. In systems that can be described analytically, a common strategy is to fit a low-order Taylor expansion of the equations of motion to the system's recent trajectory and use that short-range prediction to guide decisions, repeating the process frequently. We successfully imitated this process by tuning our agents' preferences using evolution to fit recent observations of their movement, then executing the model a short distance into the future. Like our work in JFACC, we were reasoning about geospatial movement. There, we were learning the parameters of an equation to combine information to guide the physical aircraft; in RAID, we were learning the preference vectors of individual agents. The predictions from this process were more accurate than those made by other teams using Bayesian and game-theoretic reasoners.

Second, we gave our agents a model of emotion, triggered by their environment by way of a vector of dispositions [6], and evolved this vector in response to games in which some units deliberately simulated cowardice (Chickens) or unreasonable boldness (Rambos). We were unable to evolve the dispositions for Rambos, because they usually died in combat very quickly. But we were able to identify as many Chickens as did human observers, and we identified them more quickly.

3 Introduction to SCAMP

SCAMP (Social Causality using Agents with Multiple Perspectives) is an agent-based modeling and simulation system developed for the DARPA Ground Truth program to generate realistic social data from an explicit causal model. Social scientists then used this data to test methods for causal analysis. Our model, inspired by Syria's recent history, included different groups of agents representing an oppressive government, its military, a democratic armed opposition, violent extremists, relief agencies, and civilians. We have described SCAMP at CSS 2020 [11] and elsewhere [12, 13].

SCAMP is a stigmergic system: the agents move through their environment, marking where they have been (like ants leaving pheromones) and guiding their decisions by the marks left by other agents. Their environments are directed graphs over which agents can move: a causal event graph (CEG) whose nodes are different types of events and whose edges indicate that an agent currently participating in one type of event could reasonably choose to participate in another, and a geospatial lattice allowing agents to move through space. The edges in these two graphs do not in themselves fully capture our intuitions about social causality, but they do constrain the choices that agents make, and the combination of agent choice and graphical structure *is* a strong model of social causality.

Nodes in both the CEG and geospace are characterized by vectors of *features* of three types. *Wellbeing features* are intrinsic consequences of participating in an event (e.g., impact on physical health; terrain difficulty). *Urgency features* capture the impact of participation on the goals of each group. Most directly reflecting the ant model, *presence features* record the level of recent participation by agents of each group. Urgency features are computed by a hierarchical goal network (HGN) for each group. The HGN translates the degree of activity on CEG nodes that support or block its leaf-level subgoals into an estimate of the urgency of those CEG nodes for satisfying the overall network. Each agent carries a vector of *preferences* defined over the same space. To choose its next action, the agent

- computes the dot product of its preferences and the features of each alternative,
- exponentiates each dot product to yield a positive number,
- raises each to the power of a determinism parameter,
- and normalizes them to form a roulette.

The determinism parameter allows the modeler to simulate different levels of deterministic choice. Determinism less than 1 models increasingly random behavior (with 0 yielding a complete random choice), while determinism greater than 1 models increasingly rigid preference for the strongest preference.

SCAMP's agents use the polyagent model [9]. Each agent is represented by an *avatar*, which sends out repeated waves of *ghosts* to explore possible futures. The ghosts consider the full feature vector of each alternative, and deposit the pheromones of its avatar's group on the nodes that it visits. Avatars make their decisions by constructing their roulette using only the presence features left by their ghosts.

SCAMP offers several artifacts that can readily be used as chromosomes in an evolutionary approach.

The work reported here focuses on agents' preference vectors. This was also the focus of our work in RAID and JFACC, but there our agents were constrained to a regular lattice modeling geospace. SCAMP's CEG is much more complex and irregular, and it was not clear that we could successfully learn the preferences of agents based on the sequence of behaviors that they exhibit in moving over such a space.

SCAMP's HGN, which translates information about the participation on events into estimates of the urgency of those events for different groups, is more complex than a lattice, but is more regular than the CEG. We have done some preliminary experiments applying genetic programming [4] (manipulating a graph structure as the chromosome) to learn the HGN behind observed behavior.

The CEG itself might also be a candidate for modification, both of its structure (adding and removing edges among nodes) and of its features (to learn the characteristics of event types).

4 Evolving Preferences

Applying evolution to a problem requires defining the *use case* being addressed, the *experimental framework*, and the *evolutionary process* we implemented. We also discuss the *results* we achieved.

4.1 Use Case

SCAMP produces a wide range of information about the dynamics of an artificial society, but perhaps the most important is the trajectory of each agent through the space of possible behaviors represented by the CEG. The structure of the CEG represents a collection of hypotheses by the modeler about the different plausible behavioral trajectories, but the actual trajectory followed by a given agent depends as well on the agent's preference vector, which guides its choice among alternative next events. For each group, modelers define notional values for each part of the preference vector, estimating group member's preference for: the impact of participation on their individual well-being, the strategic importance of the event to each group in the model, and the recent level of participation in the event by agents of the various groups (reflecting the agent's desire to affiliate with, or avoid, specific groups).

Preferences of individual agents are sampled from their group's vector, and generate some sequence of behaviors, reflecting the dynamic consequences of the modeler's theory about the actors in the scenario. In a real application such as intelligence analysis we really want to reason in the other direction, from a sequence of observed behaviors to the underlying preferences that drive them. The structure of the model (the behavioral alternatives in the CEG, the overall structure of goals in the HGN, the geospatial context) imposes many constraints on these choices. Given all these constraints, what set of preferences would generate a particular behavioral trajectory that has been observed? The resulting preference vector could be used for short-horizon forecasting (as we did in RAID), or to advise policy makers (similar to aircraft routing in JFACC).

4.2 Experimental Framework

We want to compare the success of evolution in different CEG structures, so instead of using the human-constructed CEG from Ground Truth, we built a CEG generator whose parameters include the number of nodes, maximum node degree, maximum path length, and whether the CEG should be constrained to a tree structure or a directed acyclic graph (DAG). We also generated a synthetic population of two groups of three agents each. The feature space has six elements, two for wellbeing features, two for an event's urgency to each group's goals, and two for the recent presence of each group on the node. In the experiments reported here, we did not run the HGN, so the first four features of each node were assigned random numbers in [0, 1]. However, the last two features were modified dynamically as the simulation ran.

Our baseline consists of 300 nodes. We define an initial START node (node 0), and a single STOP node (node 301). When an agent reaches STOP, it cycles back to START and runs again, though because of the pheromones that have been deposited during its first run, it often follows a different trajectory. All nodes are accessible to both groups of agents.

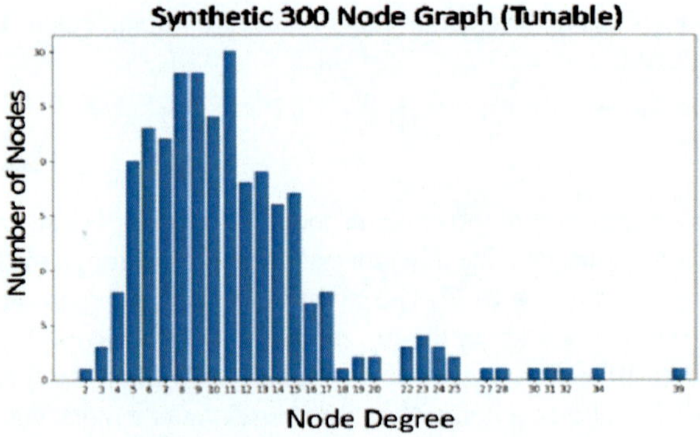

Fig. 1. Node degree distribution for 300 node synthetic CEG

Figure 1 shows the distribution of node degrees in a 300-node synthetic DAG, truncated at 40. Three nodes have higher degrees: node 80 (degree 65), node 244 (degree 66), and node 84 (degree 138). This long-tailed structure reflects the degree structure of the 460-node CEG our analysts constructed for Ground Truth (Fig. 2).

The model has two groups, G0 and G1, each with three agents. The baseline preferences for G0 are <.8, −.8, .8, −.8, .8, −.8>, and for G1, <−.8, .8, −.8, .8, −.8, .8>, and each agent's preferences are selected with variation 0.4 around its group's baseline. The last two elements in each vector mean that each group's agents are attracted to other agents of the same group, and repelled from agents of the opposite group.

Table 1 summarizes the trajectories traveled by each agent in one run. Agents that reach the end of the CEG before the end of the run start again at the beginning, generating a new trajectory. Each trajectory is a sequence of event nodes traversing the CEG. Bold trajectory numbers indicate problems we could not solve, while the bold count under each agent mark that agent's initial trajectory, details whose relevance will become clear.

The two groups show very distinctive behaviors. G0 explores a total of nine different trajectories, including all instances of trajectories of length 7 or greater, while G1 explores only three different trajectories, all distinct from those of G0. In spite of the random

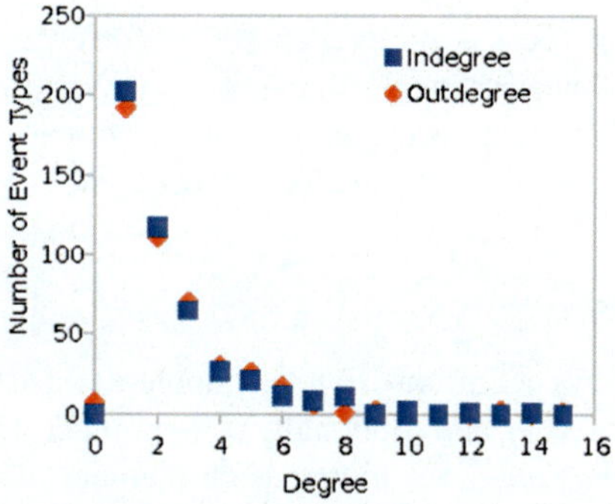

Fig. 2. Degree distribution in Ground Truth model

Table 1. Trajectories in experimental run

Trajectory #	Length	Group 0			Group 1			Total traj's
		A302	A402	A502	A612	A712	A1012	
1	6	2	3	**3**				8
2	8	4	3	1				8
3	10	1		1				2
4	11	1	2	1				4
5	10	2	2	5				9
6	4	1	3	1				5
7	7	1	1	1				3
8	5	**1**	**1**					2
9	9		1					1
10	6				**1**			1
11	4				28	**29**	30	87
12	5						**1**	1
Distinct trajectories		8	8	7	2	1	2	12
Total trajectories		13	16	13	29	29	31	131

way in which features were generated and the very schematic definition of the group preferences, this distribution of trajectories is a reasonable facsimile of two purposeful groups with very different objectives.

SCAMP is designed to be restarted from the point it terminates, and when it ends a run, it encodes its complete state in a snapshot. This snapshot records the complete trajectory of each avatar, and also the complete history of feature space over all nodes throughout the run. The latter history allows us to run avatars through the period covered by the snapshot without deploying new ghosts, simply sampling the pheromones deposited by the ghosts at each epoch during the run that generated the snapshot.

The genetic algorithm seeks a preference vector for each avatar and each of its trajectories in a given run that can generate that trajectory. We allow a separate preference vector for each trajectory, because the original agent's preference vector can change over time (specifically, its preference for the wellbeing features).

4.3 Evolutionary Process

Our evolutionary code allows definition of a fixed population size (500 in the experiments here) and mutation chance (0.1), and (optionally) one-point crossover (which we turned on). To produce a new generation, we mutate each member of the current population, and cross it over with a partner that is at least as fit. This process triples the size of the population, so we sort the resulting genomes by fitness and crop the resulting list back to the specified population size.

The fitness of a genome (a preference vector) is the similarity of the trajectory it produces with the trajectory it is trying to match. The core of our fitness function is the string edit distance between the two trajectories, the number of nodes in one trajectory that must be inserted, removed, or changed to transform it into the other (the Levenshtein distance [5]), for which an $O(mn)$-time algorithm is available [15]. Lower values indicate fitter genomes.

Because SCAMP's decision logic is stochastic (based on roulette selection), two different runs of a given genome against the reconstructed history of the feature vectors may yield different trajectories. To evaluate the fitness of a genome, we run it a specified number of times (50 for the results reported here), and report the total Levenshtein distance over the runs, divided by the number of perfect runs plus 1. Thus a fitness of 0 indicates a perfect match on every run.

The algorithm continues until the fitness in the current generation is 0.0, or until 500 generations have run.

4.4 Analysis of Results

A "problem" is one trajectory of one avatar, identified by the avatar's name followed by the run number of the trajectory. Thus the first trajectory of A301 is A301-1 (whatever the specific trajectory might happen to be), the second trajectory (which might be the same as or different from the first) is A301-2, and so forth. Our data contains 131 problems. Of these, evolution solved 122, in very few generations (Fig. 3). The worst fitness among the unsolved problems is 0.52. The 12 distinct trajectories appear multiple times, as shown in Table 1, and all but three (trajectories 8, 10, and 12) are solved in some problem. Table 2 shows the success rate by group membership of the agents.

Of the nine problems that evolution did not completely solve, seven are from group 0, and only two from group 1 (Table 2). In addition, all nine unsolved problems are in the first four trajectories traversed by their agents. Figure 4 plots both total problems and number solved as a function of run number. The three unsolved trajectories, 8, 10, and 12, all have run number 1, and account for five of the nine unsolved problems. The other four all involve initial trajectories that are solved successfully when they appear later in an agent's history. Compare the bold numbers in Table 1.

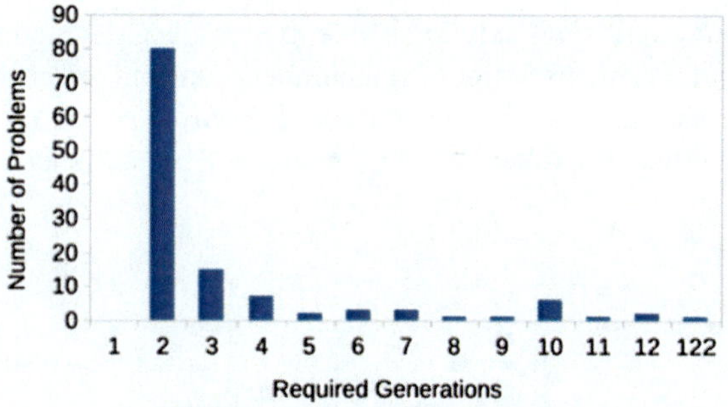

Fig. 3. Generations to solution

Table 2. Problem solution by group

Group	Total	Solved	Rate
0	42	35	0.83
1	89	87	0.98
Total	131	122	0.94

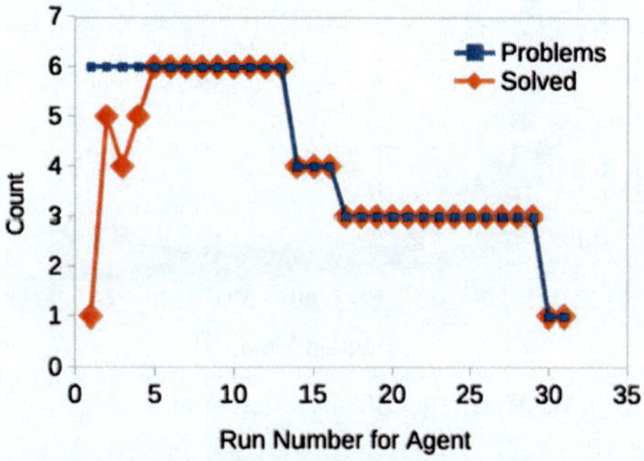

Fig. 4. Solved problems by run number

The explanation for this distribution of success and failure appears to lie in the time required for the presence pheromones to stabilize when the simulation first starts. We computed the entropy over each roulette that an avatar constructed in the original simulation, to see how much guidance was available as a function of time. Figure 5 shows the resulting pattern for group 0, and Fig. 6 for group 1. Points with 0 entropy correspond to nodes in the CEG with only a single successor, so that the roulette has only one element. In both cases, the roulette entropy varies widely during a startup period, before decreasing from its highest values and stabilizing. A high entropy roulette has segments of about the same size, meaning that the choice approaches random. The consequence of this variation is that during this start up period, the pheromone field gives agents little guidance. In terms of evolutionary learning, even if a genome has a good set of preferences, it is unlikely to be able to reproduce the required trajectory in such an environment. Problems with low run numbers confront the genome with highly variable guidance, leading to failures. In addition, the startup period for group 0 is more than twice that for group 1, explaining why the group 1 has a higher success rate than does group 0.

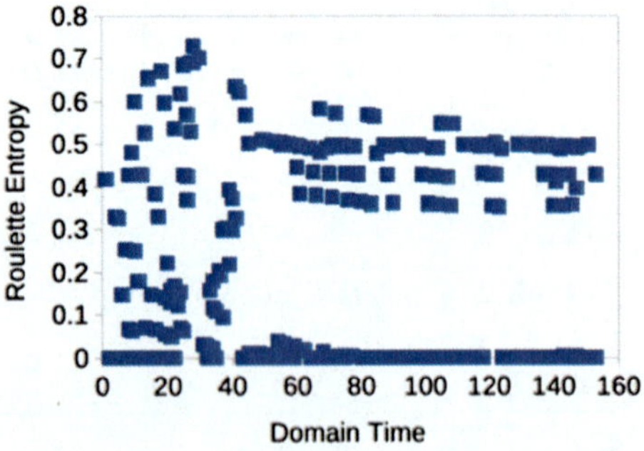

Fig. 5. Roulette entropy over time, Group 0

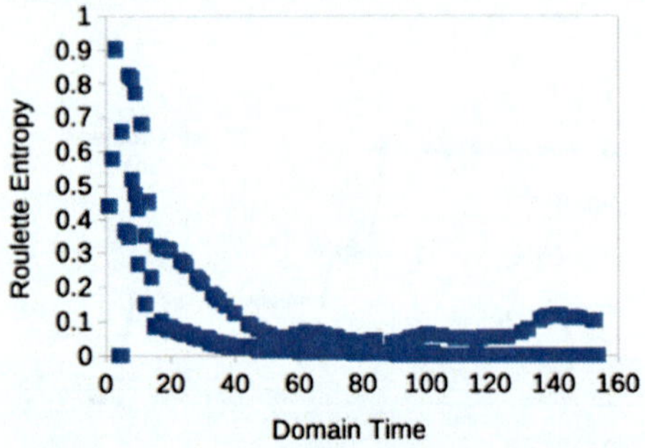

Fig. 6. Roulette entropy over time, Group 1

It is interesting to compare the roulettes evolved by the successful agents with those that generated the trajectory in the baseline run, as measured by their cosine similarity. Two perspectives on these similarities are useful.

Figure 7 shows the cosine similarity between baseline and evolved preference vectors for each distinct trajectory (a row in Table 1). Trajectories 8, 10, and 12 are missing because no evolved vector succeeded in matching the baseline trajectory. The vertical spread of similarities shows that a range of genotypes may yield the same phenotype. The preference vector is not the only constraint on agent movement. Others are presented by the structure of the CEG and the feature values on each node, and these constraints can lead to the same trajectory even for preference vectors that differ in some ways from one another.

Figure 8 plots the cosine similarities as a function of trajectory length. Thus each column combines different trajectories of the same length. Longer trajectories clearly require an evolved set of preferences to be more similar to the original, because there

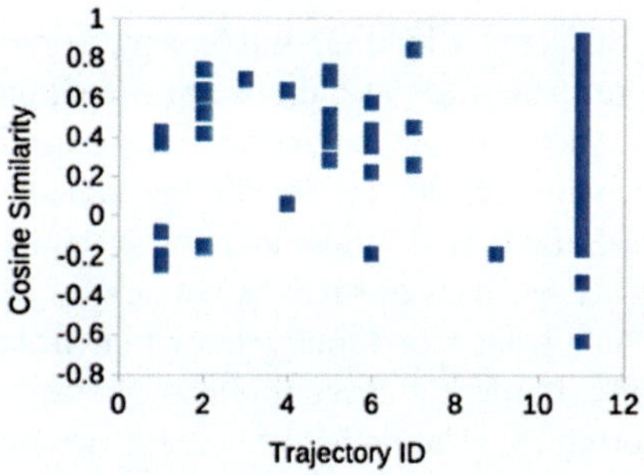

Fig. 7. Cosine similarity between different trajectories

are more decision points at which the evolving agent might be led astray by divergent preferences.

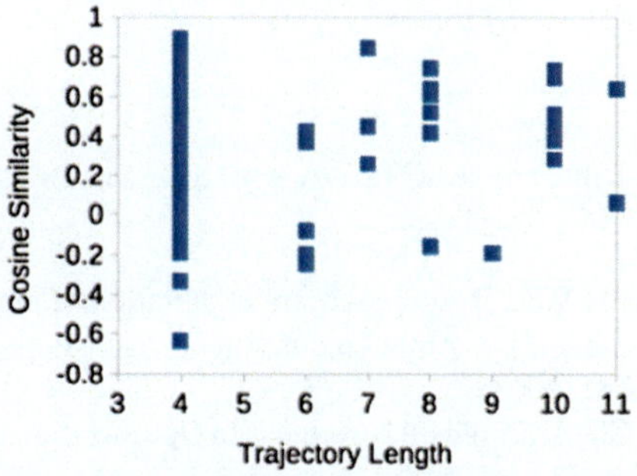

Fig. 8. Cosine similarity of original and evolved preference vectors

5 Discussion

This work leads to several lessons and directions for modeling social systems.

First, a genetic algorithm can recover the preferences of stigmergic agents moving on an irregular DAG. This capability is valuable for both planning and short-term prediction of the behaviors of actors in complex social scenarios. Particularly for short-term prediction, rapid evolution is critical to allow timely production of short-term forecasts, so the small number of generations needed (Fig. 3) is encouraging, and further work to refine this capability will explore the sensitivity of the results to population size and number of runs used to average.

Second, synthetic data to support such experiments can be generated simply, yet yield realistic patterns of behavior. We were surprised at how divergent the behaviors of the two groups were, and how the preferences of one were easier to recover than those of the other (Table 2).

Third, the pattern of success and failure in learning emphasizes the importance of understanding the stabilization period of a simulation, and illustrates how that period may vary from one group of agents to another. It is accepted practice to run a simulation for a burn-in period before gathering data; the sharp transitions in Fig. 5 and Fig. 6 suggest that automated mechanisms may be developed to optimize this period.

Fourth, the diversity of genomes that can lead to the same behavior (Fig. 7) emphasizes that SCAMP's causal dynamics, like those in living systems, pose a tension between agents' nature and nurture. Their behavior results not simply from their internal structure and motivation, nor can it be ascribed solely to their environment. The two interact, and systems like SCAMP with clearly accessible causal structures provide a promising environment for exploring the relation between these two important contributors to behavior.

Acknowledgments. The development of SCAMP was funded by the Defense Advanced Research Projects Agency (DARPA), under Cooperative Agreement HR00111820003. The content of this paper does not necessarily reflect the position or the policy of the US Government, and no official endorsement should be inferred.

References

Papers by the author are available online at https://www.abcresearch.org/abc/papers, in a browser with Javascript enabled.

1. Brueckner, S., Parunak, H.V.D.: Resource-aware exploration of emergent dynamics of simulated systems. In: Proceedings of Autonomous Agents and Multi-agent Systems (AAMAS 2003), pp. 781–788. ACM (2003)
2. Corne, D., Dorigo, M., Glover, F. (eds.): New Ideas in Optimization. McGraw-Hill, New York (1999)
3. García, A.P., Rodríguez-Patón, A.: EvoPER - an R package for applying evolutionary computation methods in the parameter estimation of individual-based models implemented in Repast. PeerJ Prepr. **4**, e2279 (2016)
4. Koza, J.R.: Genetic Programming: On the Programming of Computers by Means of Natural Selection. MIT Press, Cambridge (1992)
5. Levenshtein, V.I.: Binary codes capable of correcting deletions, insertions, and reversals. Sov. Phys. Dokl. **10**(8), 707–710 (1966)
6. Parunak, H.V.D., Bisson, R., Brueckner, S., Matthews, R., Sauter, J.: Representing dispositions and emotions in simulated combat. In: Thompson, S.G., Ghanea-Hercock, R. (eds.) DAMAS 2005, at AAMAS 2005. LNCS (LNAI), vol. 3890, pp. 51–65. Springer, Heidelberg (2006). https://doi.org/10.1007/11683704_5
7. Parunak, H.V.D., et al.: Predictive analysis for C4ISR. In: Proceedings of 25th Army Science Conference (2006)
8. Parunak, H.V.D.: Real-time agent characterization and prediction. In: International Joint Conference on Autonomous Agents and Multi-Agent Systems (AAMAS 2007), Industrial Track, pp. 1421–1428. ACM, Honolulu (2007)
9. Parunak, H.V.D., Brueckner, S.: Concurrent modeling of alternative worlds with polyagents. In: Antunes, L., Takadama, K. (eds.) MABS 2006. LNCS (LNAI), vol. 4442, pp. 128–141. Springer, Heidelberg (2007). https://doi.org/10.1007/978-3-540-76539-4_10
10. Parunak, H.V.D., Belding, T.C., Brueckner, S.A.: Prediction horizons in agent models. In: Weyns, D., Brueckner, S.A., Demazeau, Y. (eds.) EEMMAS 2007. LNCS (LNAI), vol. 5049, pp. 88–102. Springer, Heidelberg (2008). https://doi.org/10.1007/978-3-540-85029-8_7

11. Parunak, H.V.D.: Psychology from stigmergy. In: Computational Social Science (CSS 2020), CSSSA, Santa Fe, NM (2020, forthcoming)
12. Parunak, H.V.D.: Social simulation for non-hackers. In: Van Dam, K.H., Verstaevel, N. (eds.) MABS 2021. LNCS, vol. 13128, pp. 1–14. Springer, London (2021). https://doi.org/10.1007/978-3-030-94548-0_1
13. Parunak, H.V.D., Greanya, J., McCarthy, M., Morell, J.A., Nadella, S., Sappelsa, L.: SCAMP's stigmergic model of social conflict. Comput. Math. Organ. Theory (2021). https://doi.org/10.1007/s10588-021-09347-8
14. Sauter, J.A., Matthews, R., Parunak, H.V.D., Brueckner, S.: Evolving adaptive pheromone path planning mechanisms. In: Proceedings of Autonomous Agents and Multi-agent Systems (AAMAS02), pp. 434–440. ACM (2002)
15. Wagner, R.A., Fischer, M.J.: The string-to-string correction problem. J. ACM **21**(1), 168–173 (1974)

Drivers and Predictors of COVID-19 Vaccine Hesitancy in Virginia

Asal Pilehvari[1](✉), Jason Ton[2], Mukundan Ram Mohan[2], Achla Marathe[1,2], and Anil Vullikanti[2]

[1] Department of Public Health Sciences, School of Medicine, University of Virginia, University of Virginia Cancer Center, Charlottesville, USA
asal@virginia.edu

[2] Network Systems Science and Advanced Computing, Biocomplexity Institute, University of Virginia, Charlottesville, USA

Abstract. This research uses the COVID-19 Trends and Impact Survey provided by Carnegie Mellon University in partnership with Facebook to study predictors and drivers of COVID-19 vaccine hesitancy in Virginia's adult population. It estimates vaccine hesitancy rates at a zip code level in Virginia by applying multilevel statistical models. Our analysis identifies the demographic features of zip codes that are associated with vaccine hesitancy. It also examines the drivers of COVID-19 vaccine hesitancy across Virginia. Results show the presence of a larger percentage of Black and White population and a lower percentage of Hispanic population are predictors of higher vaccine hesitancy within a zip code in Virginia. Among these drivers, the biggest is system distrust, where individuals either do not trust the government or believe that the vaccine is not efficacious. Finally, it provides policy insights and tailored outreach programs for improving COVID-19 vaccination acceptability in different regions in Virginia.

Keywords: COVID-19 · Vaccine hesitancy · Distrust · Vaccine side effects · Multilevel analysis · CTIS survey · Virginia

1 Introduction

The novel coronavirus disease (COVID-19) has caused more than 4 million deaths worldwide as of July 2021, the largest in human history. Although COVID-19 preventive behaviors such as mask-wearing and social distancing have been shown to be effective in curbing the spread of the virus [1, 2], effective control of the pandemic depends on the development and uptake of vaccines, which have been proven to be very effective. To achieve herd immunity, at least 60–70% of the population needs to be fully vaccinated against COVID-19. [3] Since March 2021, several vaccines have become available in the US, [4] and over 68% of the adult population has been vaccinated by end of July 2021. However, the virus strain is evolving and the vaccinated population is not uniformly distributed in the US. The Southern States have much smaller vaccination rates compared to the Northern States. As of now in July 2021, the Delta variant of the COVID-19 virus

Z. Yang and E. von Briesen (Eds.): CSSSA 2021, SPCOM, pp. 98–111, 2022.
https://doi.org/10.1007/978-3-030-96188-6_8

is spreading rampantly in States like Florida and Texas where less than 50% of people are fully vaccinated. The uneven distribution of the unvaccinated population keeps the US from achieving herd immunity, even when overall vaccination rates may be fairly high in many regions. According to the World Health Organization (WHO), vaccine hesitancy is defined as "*delay in acceptance or refusal of safe vaccines despite availability of vaccination service*" and is influenced by several factors including issues of confidence (do not trust vaccine or provider), complacency (do not perceive a need for a vaccine, do not value the vaccine), and convenience (access) [5].

Vaccine hesitancy is not a new issue in the U.S— the most recent example is the recurrence of measles in the last few years due to falling immunization rates. [6] It has been found that non-compliance is often spatially clustered, and this can lead to a significant risk of outbreaks, even if the overall vaccination rate is above the herd immunity level. [7] With regard to the COVID-19 vaccine, studies found the vaccine acceptance varied vastly across different regions in the US even before the vaccine became available. Early in the pandemic, a representative survey of over 5000 American adults revealed that more than one-third of adults would refuse to get vaccinated. [8] As of November 2020, Raja et al. [9] found that 59.1% of adults between 18–88 years old in a cross-national survey were either vaccine-hesitant or unsure about accepting the COVID-19 vaccine in the US. Later in late 2020 and early 2021, using data of 37,470 nationwide app users, McCabe et al. [10] found majority of people were undecided about taking the vaccine and 19% were vaccine-hesitant.

A recent study reports that COVID-19 vaccine hesitancy rates vary between 26–44% in the US. [11] Particularly, researchers found the vaccine hesitancy was more severe in parts of Virginia, West Virginia, and Georgia.[1] According to the US Center for Disease Control and Prevention (CDC), by the end of March 2021, only 17.1% of Virginians were fully vaccinated, this reached 50% by late-June, 2021 as those who wanted the vaccine likely got vaccinated. [12] However, as the supply of vaccines increased, the demand dropped to a rate close to the national rate. Studies based on non-representative samples found the willingness to take the COVID-19 vaccine was associated with age, gender, race, ethnicity, income, and education levels. [13, 14] Identifying individuals who do not want to take the vaccine, and the reasons behind it, is critical for officials for designing effective vaccine uptake programs.

The reasons for hesitancy in vaccine uptake are complex and vary geographically, [15] but the key hurdles anecdotally appear to be centered around the possibility of side effects and doubts about vaccine safety given the short timeline of vaccine development, testing, and trials. Although the COVID-19 vaccines are not yet approved by the US Food and Drug Administration (FDA), they are proving to be highly efficacious in preventing deaths and hospitalizations. The latest Delta variant is twice as infectious as the original virus and has caused over 80% of the current COVID-19 infections and is responsible for over 90% of deaths and hospitalizations in the US. [16] Despite this, over 30% of all US adults have not had a single dose of the COVID-19 vaccines. In order to persuade the unvaccinated to get the vaccine, we need to understand the drivers and predictors of vaccine hesitancy. This research aims to do just that through the following:

[1] https://thefulcrum.us/big-picture/vaccine-hesitancy.

1) estimate the vaccine hesitancy rates at the zip code level in Virginia using individual-level data, 2) identify the demographic features of zip codes that are associated with vaccine hesitancy, 3) examine the reasons behind the refusal of COVID-19 vaccine across Virginia, 4) provide policy insights and tailored outreach programs for improving COVID-19 vaccination acceptability in different regions in VA.

Our results show that the presence of a larger percentage of Black and White population and a lower percentage of Hispanic population are predictors of higher vaccine hesitancy within a zip code in Virginia. A larger share of the younger population share (<20 years old) is significantly associated with higher levels of vaccine hesitancy. Among the drivers of vaccine hesitancy, the biggest driver is system distrust, where individuals either do not trust the government or believe that the vaccine is not efficacious. This is followed by the fear of side effects from the vaccine. However, the cost of the vaccine and the religious beliefs are not significant drivers of vaccine hesitancy in Virginia.

2 Data

2.1 Vaccine Hesitancy

We use the COVID-19 Trends and Impact Survey (CTIS) provided by Carnegie Mellon University (CMU) in partnership with Facebook. The CTIS is a daily cross-sectional survey that encourages Facebook's adult users to take the survey. The US version of the survey is run by CMU and the global version by the University of Maryland, both in partnership with Facebook. [17] CTIS data provides respondents' information such as gender, age range, education, and location by zip code. Due to confidentiality rules, information about the race and ethnicity of respondents is not available in the survey. However, we obtain zip code level aggregate data on demographics through Zip Code Tabulation Areas (ZCTA) for the analysis in this study. [18] Survey questions regarding vaccines were added to the survey starting January 2021, and detailed hesitancy questions were added later in early February. We apply the CTIS data that were collected in March 2021 to understand the reasons for hesitation in vaccine uptake. CTIS collected data for 34,877 adults who represented 805 zip codes in Virginia. CTIS respondents represented 6,163,482 Virginian adults after incorporating the survey weights. [19] Intent to be vaccinated against COVID-19 was measured with the question, "If a vaccine to prevent COVID-19 were offered to you today, would you choose to get vaccinated?" Response options were "yes, definitely," "yes, probably," "no, definitely not," and "no, probably not." Participants who responded "no, definitely not" or "no, probably not" are considered vaccine-hesitant in this study.

2.2 Vaccine Hesitancy Drivers

Those who responded "no, definitely not" or "no, probably not" were asked to provide the reason for their unwillingness to take the vaccine and get immunized against COVID-19. Response options included a wide range of reasons from being concerned about possible side effects to religious reasons. We categorized the hesitancy reasons into six major categories: 1) Watchful: those who are waiting to see if the vaccine is safe on others and may choose to get it later, 2) Side-effects: those who are concerned about the possible side effects of the vaccine or having allergic reactions, 3) System-distrustful: individuals who do not have trust either in the government or in the efficacy of vaccines, 4) Health-conditions: this category includes three types of reasons: those who have specific health conditions and are concerned about the safety; those who are pregnant or want to get pregnant; and those whose doctor has not recommended it, 5) Cost-Anxious: people with concern about the cost of the vaccine, and 6) Religion: those who deem vaccine is against their religious beliefs.

2.3 Demographics/Predictors

US Census provides the Zip Code Tabulation Areas (ZCTA) to overcome the difficulties in defining the land area encompassed by a ZIP code in the US. ZCTAs are created through the aggregation of Census blocks into areas that most closely correspond with zip code areas which contain population demographic composition within zip codes. [18] In our analysis, we use the latest available ZCTA that is used in the 2010 Census. In particular, available information in ZCTA includes population and population share by gender, age, race, and the number of housing units within a zip code.

3 Method

Given the two-level hierarchical structure of the data—individuals nested within zip codes, multilevel statistical models are ideal methods to estimate: 1) vaccine hesitancy intensity within zip codes while controlling for the individual-level heterogeneity, 2) estimating the association of zip code demographic composition with the intensity of vaccine hesitancy within zip codes.

In particular, two-level hierarchical generalized linear models are implemented using individual-level characteristics at level-1 and zip code demographic features as contextual-level variables at level-2. The first model is a 2-level model that measures the effect of individual-level characteristics on vaccine hesitancy and estimates the vaccine hesitancy intensity of each zip code. The second model is a level-2 predictor for an intercept-as-output model that measures the effect of zip code demographic composition on vaccine hesitancy. In other words, the predictors of vaccine hesitancy are determined in the second model. Details of each model are provided in the next section. Interested readers can find a more comprehensive review of multilevel models in Goldstein and Raudenbush [20].

3.1 Multilevel Analysis: Identifying Vaccine-Hesitant Zip Codes

The level-1 model that only includes individual-level variables is defined as follows:

$$y_{ij} = \beta_{0j} + \sum_{p} \beta_p X_{pij} + \varepsilon_{ij} \tag{1}$$

where y_{ij} indicates the vaccine hesitancy of individual i within zip code j; it is 1 if individual i is hesitant to get vaccinated against COVID-19 and 0 otherwise. β_{0j} is the intercept. X_{pij} indicates the value of individual-level covariate, p represents demographics such as gender, age, or education. ε_{ij} is the individual-specific error component with zero mean. The level-2 model in which β_{0j} appears as output is defined as:

$$\beta_{0j} = \gamma_{00} + u_{0j} \tag{2}$$

where γ_{00} is zip code vaccine hesitancy mean, and u_{0j} is the random effect within zip codes, which is assumed to be normally distributed with mean zero and variance τ_{00}. u_{0j} is the zip code-specific error component that is shared between individuals of the same zip code (called random intercept) and represents the combined effects of omitted zip code characteristics or unobserved heterogeneity at the zip code level (i.e., access to vaccine center, specific dissemination strategies, etc.). If u_{0j} is positive, the total residual for zip code j will tend to be positive, leading to a higher probability of vaccine hesitancy than predicted by the covariates. Because u_{0j} is shared by all individuals located in the same zip code, it carries within zip code dependence among individuals. Therefore, estimating u_{0j} in level-2 allows the intercept to vary for different zip codes; positive u_{0j} estimations can be interpreted as a higher probability of vaccine hesitancy in zip code j compared to the average hesitancy among the zip codes in Virginia, and negative u_{0j} estimation implies a lower vaccine hesitancy probability than the average hesitancy of zip codes in Virginia. Accordingly, we interpret zip codes with positive u_{0j} as vaccine-hesitant zip codes.

3.2 Level-2 Predictor for Intercept-As-Output: Identifying the Predictors of the Vaccine Hesitancy

The level-2 predictor for intercept-as-output model identifies the predictors of vaccine hesitancy in a zip code, controlling for individual-level observable heterogeneities. In this model, the level-1 equation is the same as the Eq. (1) in the previous model but the zip code demographic features are added to the level-2 equation as follows:

$$\beta_{0j} = \gamma_{00} + \sum\nolimits_{q} \beta_q Demo_{qij} + v_{0j} \tag{3}$$

where $\sum_q \beta_q Demo_{qij} + v_{0j}$ equals the estimated random residuals in Eq. (2). $Demo_{qij}$ indicates demographic characteristics, q, of the zip code that can be a percentage of population by gender, race, age, or a number of housing units. v_{0j} is the i.i.d. error term.

4 Results

4.1 Summary Statistics

Table 1 reports the descriptive statistics of the demographic variables within the subsample of vaccine-hesitant and non-hesitant individuals in the CTIS survey. As shown in Table 1, the majority of individuals are female in both subsamples (65.5% in the non-hesitant and 62.3% in the hesitant subsample) but the non-hesitant group contains significantly more females than the hesitant group. Age composition of the non-hesitant subgroup shows 37.4% are between 45–64 years old, 31.9% are older than 65 years, and 30.7% are younger than 45 years. Whereas in the vaccine-hesitant subsample, 38.4% are between 45–64 years old, only 14.3% are 65 or above, and the majority are younger than 45 years old (47.4%).

Table 1. Demographic characteristics of CTIS Survey sample by vaccine hesitancy

	Subsample of vaccine		Difference
	Non-hesitant[a]	Hesitant[b]	
	Mean	Mean	
Female	0.655	0.623	0.032***
	(0.475)	(0.485)	(0.008)
Age group			
<45	0.307	0.474	−0.167***
	(0.461)	(0.499)	(0.008)
45–64	0.374	0.384	−0.010
	(0.484)	(0.486)	(0.008)
>65	0.319	0.143	0.176***
	(0.466)	(0.350)	(0.008)
Education group			
High school or less	0.147	0.299	−0.151***
	(0.355)	(0.458)	(0.006)
Some college and degrees	0.253	0.114	0.139***
	(0.435)	(0.317)	(0.007)
Master's degree or higher education	0.600	0.588	0.012
	(0.490)	(0.492)	(0.008)
Observations	29,717	5,160	

Note: This table shows the average of demographic variables within the hesitant and non-hesitant subsample of individuals in the CTIS survey. [a] The non-hesitant subsample refers to the CTIS respondents that answered "yes, definitely," "yes, probably," to the question regarding the intention to get vaccinated against COVID-19. [b] Vaccine hesitant subsample refers to the CTIS respondents that answered "no, definitely not," "no, probably not," to the question regarding the intention to get vaccinated against COVID-19. Standard deviations are in the parenthesis. Asterisks show the difference between the hesitant and non-hesitant subsamples is statistically significant as follows: ***at 1% level; ** at 5% level, and * at 10% level.

Statistically, a significant difference exists between the two subsamples in the share of individuals younger than 45 and 65+ years old but not in the age group of 45–64 years old. In both subsamples, individuals with "master's degree or higher education levels" form the largest group; these are followed by individuals with "some college degree" in the non-hesitant subsample, and "High school or less education level" in the vaccine-hesitant subsample. The portion of "High school or less educated" individuals is significantly larger in the vaccine-hesitant subsample than the non-hesitant subsample (29.9% versus 14.7%), while the portion of individuals with "some college degree" is significantly smaller in the vaccine-hesitant subsample than the non-hesitant subsample. In summary, Table 1 shows that vaccine hesitancy is higher among <45 years old individuals and the less educated, but lower among 65+ years old and college-educated.

These systematic differences between the profile of individuals in vaccine-hesitant versus non-hesitant sub-groups highlight the importance of controlling for individual heterogeneity in the estimation of zip codes vaccination hesitancy levels. Multilevel methods enable us to use individual/resident level information available in CTIS data to estimate vaccine hesitancy at the zip code level while controlling for heterogeneities.[2]

4.2 Identifying Vaccine-Hesitant Zip Codes

Vaccine hesitancy estimation within zip codes using 2-level analysis (Eqs. (1) and (2)) is displayed in Fig. 1. According to this Figure, at least a few zip codes suffer from high rates of vaccine hesitancy in all the Virginia regions, except the Northern region where a majority of zip codes are non-hesitant.

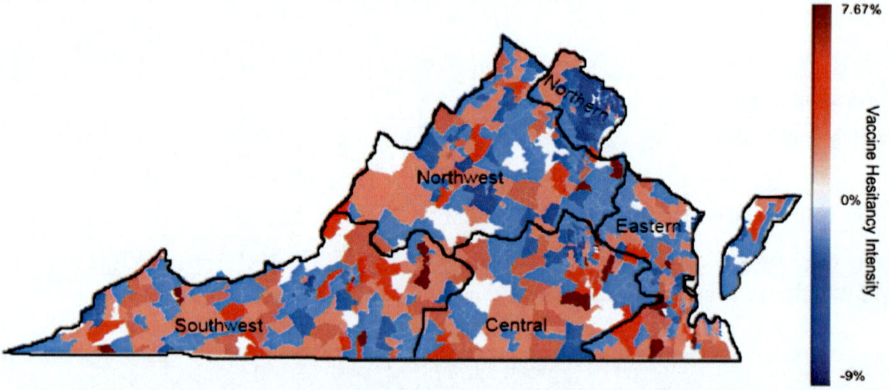

Fig. 1. Vaccine hesitancy intensity by zip codes in Virginia

Zip codes are labeled as vaccine-hesitant or non-hesitant, based on estimates of vaccine hesitancy intensity using the multilevel analysis (Eq. (1) and (2)). The zip codes with positive estimates of vaccine hesitancy are labeled as the vaccine-hesitant zip codes (353 zip codes), and zip codes with negative intensity estimates are labeled as the non-vaccine-hesitant zip codes (452 zip codes).

Predictors of Vaccine Hesitancy. The average effect of each demographic characteristic on the intensity of vaccine hesitancy of zip codes, while controlling for the other

[2] Use of aggregate methods such as "scan statistics" for identifying areas with high prevalence of a phenomena (e.g., vaccine hesitancy in this study) results in a biased estimation since they ignore the heterogeneity among individuals within that area.

characteristics, is provided in Table 2. As is shown in Table 2, a larger share of Black and White population within a zip code significantly increases the vaccine hesitancy within the zip code, holding other variables fixed. This finding is in line with the findings in other studies on vaccine hesitancy of either COVID-19 [14] or other diseases such as flu. [21] Share of Hispanic population is negatively correlated with the intensity of hesitancy against COVID-19 vaccines. As compared to the share of the younger population (<20 years old), a larger population in the age range of 20–64 years old or even older than 65 years old is associated with lower vaccine hesitancy in a zip code. In other words, a larger share of the younger population (<20 years old) is significantly associated with higher vaccine hesitancy. Zip codes with more housing units experience marginally lower vaccine hesitancy.

Table 2. Estimation of demographic factors' impact on intensity of vaccine hesitancy in Virginia zip codes.

	Intensity of vaccine hesitancy[a]
Male Population (%)	0.009
	(0.01)
White Population (%)	0.116***
	(0.02)
Black Population (%)	0.143***
	(0.02)
Hispanic Population (%)	−0.035**
	(0.02)
Age group	
20–64	−3.631***
	(1.34)
>65	−0.027**
	(0.01)
Number of housing units/1000	−0.034*
	(0.02)
Constant	−8.722***
	(1.77)
Observations	805

Note: This table presents the multiple regression estimation results of vaccine hesitancy intensity on demographic characteristics of zip codes. [a] Intensity of vaccine hesitancy within zip codes is obtained by a 2-level hierarchical estimation of Eq. (1) and Eq. (2). Standard errors are provided in the parenthesis. Asterisks show the difference between the hesitant and non-hesitant subsamples is statistically significant as follows: *** at 1% level; ** at 5% level, and * at 10% level.

4.3 Vaccine Hesitancy Drivers

Now that we have identified vaccine-hesitant zip codes in Virginia and the potential predictors, it is critical to find out the drivers of hesitancy in order to address it through tailored interventions. In both vaccine-hesitant and non-hesitant zip codes, there are individuals who are vaccine-hesitant. Table 3 shows the drivers of vaccine hesitancy among the hesitant population in both types of zip codes. As is shown, on average 35.4% of the population in vaccine-hesitant zip codes (whereas only 5.7% in the non-hesitant zip codes) is hesitant to get vaccinated against COVID-19, which is similar to the national rate of hesitant population [22].

Table 3. Summary statistics of vaccine hesitancy drivers in Virginia zip codes

	Non-hesitant zip codes	Vaccine hesitant zip codes[a]
Vaccine hesitant population (%)	0.0577	0.354
	(0.0613)	(0.186)
Hesitancy drivers		
Watchful (%)[b]	41.90	37.15
	(31.86)	(29.91)
System distrustful (%)[c]	46.83	55.99
	(32.63)	(30.98)
Side effects (%)[d]	54.84	52.24
	(32.06)	(30.90)
Health conditions (%)[e]	25.28	23.85
	(28.93)	(27.05)
Religious (%)[f]	5.330	6.805
	(16.03)	(16.69)
Cost-anxious (%)[g]	3.480	3.299
	(10.89)	(8.850)

Note: This table provides the percentage of the hesitant population in both types of zip codes, as well as its distribution by hesitancy drivers. Vaccine hesitant zip codes refer to zip codes with positively estimated vaccine hesitancy value in Eq. (1) (positive intercept). [b] Watchful refers to individuals who are waiting to see if the vaccine is safe and may get it later. [c] System-distrustful refers to individuals who do not have trust either in the government or vaccine efficacy. [d] Side-effects refers to those who are concerned about the possible side effects of the vaccine or have allergic reactions. [e] Health-conditions include three types of reasons: those who have specific health conditions and are concerned about the safety; those who are pregnant or want to get pregnant; and those whose doctor has not recommended it. [f] Religious indicates those who deem vaccine is against their religious beliefs. [g] Cost-Anxious refers to people with concern about the cost of the vaccine. Standard deviations are provided in the parenthesis.

Among these hesitant individuals, the majority (55.99%) are those who either do not have trust in the government or the efficacy of the vaccine. [13] The second largest hesitancy driver is the concern about the potential side effects of vaccines or allergic reactions (52.24%, on average). On average, 37.15% of the hesitant population are "watchful" meaning that they are waiting to see if the vaccine is safe and they may get it later. Specific health conditions are a driver for only 23.85% of the hesitant population, on average. Religious reasons and concerns about the cost of the vaccine account for less than 10% of the hesitant population. Therefore, the main vaccine hesitancy drivers in Virginia are system-distrust and the fear of side-effects followed by watchfulness.

5 Discussion and Conclusions

To our knowledge, this is the first state-level study based on a large representative sample where participants provide reasons for hesitancy in the uptake of the COVID-19 vaccine. Vaccine hesitancy is a complex problem [23] and numerous studies have shown that a one-size-fits-all approach will not work and the most successful vaccine programs to addressing vaccine hesitancy are those tailored to the subgroups. [24] The findings reported here, highlight regions in Virginia with subgroups who are more likely to be vaccine-hesitant and require targeted interventions to be effective.

Our results show that larger percentages of the Black and White populations are significantly associated with higher levels of vaccine hesitancy within a zip code. Association between the Black population and vaccine hesitancy may reflect the general mistrust of the Black community in the medical system. [25] Since COVID-19 has disproportionately impacted minorities, especially Black communities, [26] developing strategies to address vaccine hesitancy concerns in Black communities to increase their vaccination uptake is critical in achieving herd immunity. In fact, health inequities will likely become worse if targeted steps are not taken to engage subgroups who are most affected by COVID-19 [26] and are vaccine-hesitant.

We also observe a significant positive association between individuals who are younger than 20 years old and vaccine hesitancy in Virginia. This might be caused by parents; children younger than 18 years old need their parent's consent to get vaccination nearly in all states including Virginia (41 states).[3] Recent surveys by the Kaiser Family Foundation report a quarter of respondents would not allow their children to get vaccinated.[4] Therefore, vaccine campaigns inside zip codes with a higher teen population should engage community leaders, teachers, and parents, and focus on spreading information on the safety and efficacy of the COVID-19 vaccine for teens.

To design tailored vaccine delivery to regions and communities, it is essential to understand community-specific attitudes and perceptions towards vaccination. Our analysis shows that over half the vaccine-hesitant Virginians are concerned about vaccine side effects; this calls for vaccination campaigns on assuring people on the safety of the vaccine, particularly in Northwest and Southwest regions (red circles in Fig. 2(a)). Figure 2(b) shows that system distrust is the other dominant hesitancy driver across the

[3] https://www.kff.org/policy-watch/covid-19-vaccination-and-parental-consent/.

[4] https://www.kff.org/coronavirus-covid-19/poll-finding/kff-covid-19-vaccine-monitor-april-2021/.

state. "System distrust" reflects distrust either in the government or in the efficacy of vaccines. To overcome distrust, vaccine campaigns could show government officials, community leaders, and health regulators taking the vaccine, and could emphasize the stark differences in hospitalizations and death rates between the vaccinated and unvaccinated populations.

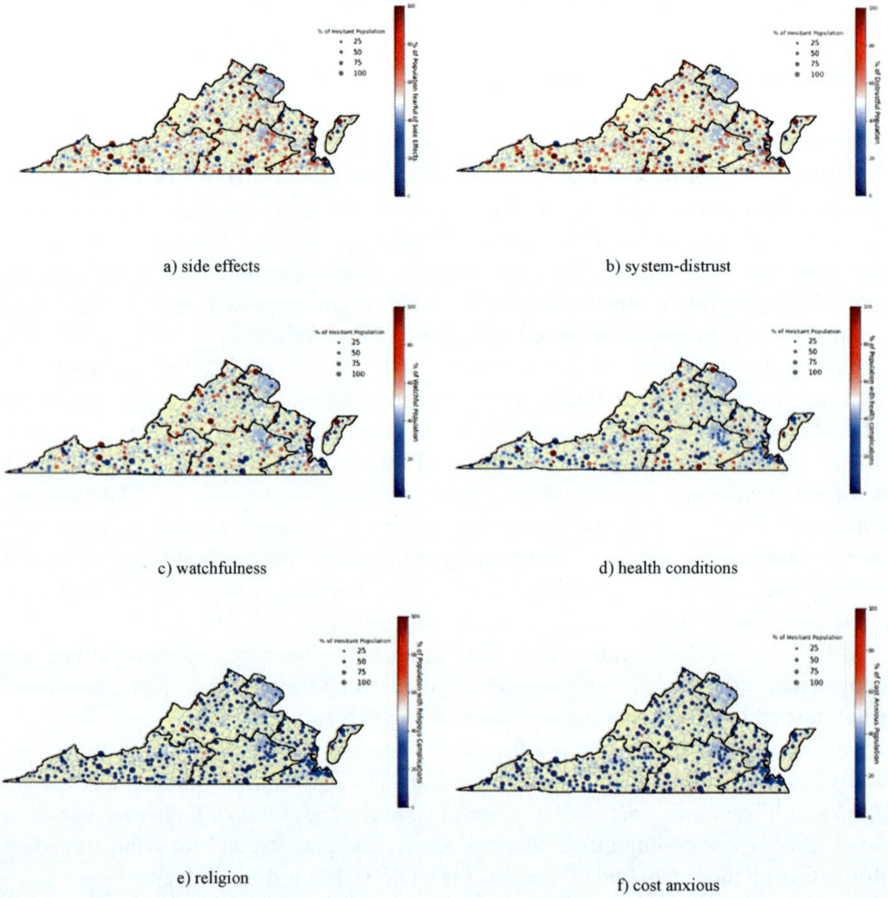

Fig. 2. Spread of vaccine hesitancy by hesitancy drivers in vaccine-hesitant zip codes across Virginia

The third major driver is "watchfulness" in which people are waiting to see the efficacy of the vaccines first. Zip codes with a larger percentage of "watchful" individuals are located in Northern, Northwest, and Southwest Virginia, as shown in Fig. 2(c). Zip codes that are hesitant due to health conditions, comorbidities, and pregnancy-related reasons, (on average 23.85% of the hesitant population) are largely located in Southwest, Northern, and Central Virginia (Fig. 2(d)). Vaccine campaign programs in these regions could make physicians or physician assistants available at vaccine sites to help answer any questions related to the safety of the vaccines for specific health conditions. A few zip codes in the Northwest suffer from vaccine hesitancy due to religious beliefs (Fig. 2(e)). In these zip codes vaccine campaigns could arrange the presence of religious leaders at vaccination sites to overcome the religion-driven hesitancy. As Fig. 2(f) depicts vaccination cost is not a significant driver of vaccine hesitancy in Virginia.

Our study has several limitations. The survey is web-based so non-participation may have introduced bias into the results, although we followed the recommended practice for this data set and weighted the results to account for non-response. [27] Secondly, the race and ethnicity of respondents are not available in CTIS data, which prevents the detailed (micro-level) estimation of predictors of vaccine hesitancy or uptake. The data used here on race/ethnicity is taken from ZCTA which provides aggregated information at a zip code level. This analysis highlights the associations between vaccine hesitancy and socio-demographic factors within zip codes. The cross-sectional nature of CTIS data prevents us to do causal inference analysis and investigate any potential causality.

Despite these limitations, this study provides insights into possible future directions for improving vaccine acceptance by developing community-based interventions in Virginia. The subgroups that have been identified as vaccine-hesitant should be targeted for interventions and their dominant hesitancy drivers should be used in designing the appropriate programs. Prior work has shown that targeted vaccination, e.g., based on the number of contacts an individual has, leads to significantly better disease outcomes. [28] Since the contact structure has significant correlations with demographics, our results can also help in such vaccine prioritization questions. Finally, our results show that vaccination programs in Virginia should emphasize the efficacy of the vaccines [29] in their messaging to help overcome the widespread perception of poor efficacy and system distrust among Virginians.

References

1. Pilehvari, A., You, W., Chen, Krulick, J., John, L., Venkatramanan, S., Marathe, A.: Differential impact of social distancing on COVID-19 spread in the U.S.: by rurality and social vulnerability. Work Pap. (2021)
2. Zhang, K., Vilches, T.N., Tariq, M., Galvani, A.P., Moghadas, S.M.: The impact of mask-wearing and shelter-in-place on COVID-19 outbreaks in the United States. Int. J. Infect Dis. **101**, 334–341 (2020). https://doi.org/10.1016/J.IJID.2020.10.002
3. Aschwanden, C.: The false promise of herd immunity for COVID-19. Nature **587**(7832), 26–28 (2020). https://doi.org/10.1038/d41586-020-02948-4
4. CDC: Different COVID-19 Vaccines | CDC. 06 April. https://www.cdc.gov/coronavirus/2019-ncov/vaccines/different-vaccines.html. Published 2021. Accessed 13 July 2021
5. MacDonald, N.E.: Vaccine hesitancy: definition, scope, and determinants. Vaccine **33**(34), 4161–4164 (2015). https://doi.org/10.1016/J.VACCINE.2015.04.036
6. McAteer, J., Yildirim, I., Chahroudi, A.: Committee for the S for PRA, Chahroudi A, Committee for the S for PRA. The VACCINES act: deciphering vaccine hesitancy in the time of COVID-19. Clin. Infect Dis. **71**(15), 703–705 (2020). https://doi.org/10.1093/cid/ciaa433
7. Cadena, J., Marathe, A., Vullikanti, A.: Finding spatial clusters susceptible to epidemic outbreaks due to undervaccination. In: Proceedings of the International Joint Conference on Autonomous Agents and Multiagent Systems: AAMAS. International Joint Conference on Autonomous Agents and Multiagent Systems, vol. 2020, pp. 1786–1788 (2020). http://www.ncbi.nlm.nih.gov/pubmed/34305484. Accessed 13 Aug 2021

8. Callaghan, T., et al.: Correlates and disparities of COVID-19 vaccine hesitancy. SSRN Electron. J. (2020). https://doi.org/10.2139/ssrn.3667971
9. Raja, A.S., Niforatos, J.D., Anaya, N., Rodriguez, R.M.: public: a cross-sectional survey 2 3. medRxiv (2021). https://doi.org/10.1101/2021.02.28.21252610
10. McCabe, S.D., et al.: Unraveling attributes of COVID-19 vaccine hesitancy in the U.S.: a large nationwide study. medRxiv Prepr Serv. Heal Sci. (2021). https://doi.org/10.1101/2021.04.05.21254918
11. Szilagyi, P.G., et al.: National trends in the US Public's likelihood of getting a COVID-19 Vaccine-April 1 to December 8, 2020. JAMA **325**(4) (2020). https://doi.org/10.1001/jama.2020.26419
12. Vaccination rates us–Google Search. https://www.google.com/search?q=vaccination+rates+us+&client=safari&rls=en&sxsrf=ALeKk00nnHarcZUCkGEPsTvu0offvzroUA%3A1627498540898&ei=LKgBYYydNsPn-gSC_4iwAw&oq=vaccination+rates+us+&gs_lcp=Cgdnd3Mtd2l6EAMyBAgjECcyAggAMgIIADICCAAyAggAMgIIADICCAAyAggAMgIIA. Accessed 28 July 2021
13. Paul, E., Steptoe, A., Fancourt, D.: Attitudes towards vaccines and intention to vaccinate against COVID-19: implications for public health communications. Lancet Reg. Heal - Eur. **1**, 100012 (2021). https://doi.org/10.1016/J.LANEPE.2020.100012
14. Malik, A.A., McFadden, S.M., Elharake, J., Omer, S.B.: Determinants of COVID-19 vaccine acceptance in the US. EClinicalMedicine **26**, 100495 (2020). https://doi.org/10.1016/j.eclinm.2020.100495
15. Robertson, E., Reeve, K.S., Niedzwiedz, C.L., et al.: Predictors of COVID-19 vaccine hesitancy in the UK household longitudinal study. Brain Behav. Immun. **94**, 41–50 (2021). https://doi.org/10.1016/J.BBI.2021.03.008
16. COVID Data Tracker Weekly Review | CDC. https://www.cdc.gov/coronavirus/2019-ncov/covid-data/covidview/index.html. Accessed 28 July 2021
17. Delphi Group: COVID-19 Trends and Impact Survey. https://cmu-delphi.github.io/delphi-epidata/symptom-survey/
18. United States Census Bureau: ZIP code tabulation areas (ZCTAs). https://www.census.gov/programs-surveys/geography/guidance/geo-areas/zctas.html
19. Barkay, N., et al.: Weights and Methodology Brief for the COVID-19 Symptom Survey by the University of Maryland and Carnegie Mellon University, in partnership with Facebook. (September 2020). http://arxiv.org/abs/2009.14675. Accessed 8 Aug 2021
20. Goldstein, H., Bryk, A.S., Raudenbush, S.W.: Hierarchical Linear Models: Applications and Data Analysis Methods. J Am Stat Assoc. **88**(421), 386 (1993). https://doi.org/10.2307/2290750
21. Quinn, S.C., Jamison, A.M., An, J., Hancock, G.R., Freimuth, V.S.: Measuring vaccine hesitancy, confidence, trust and flu vaccine uptake: results of a national survey of white and African American adults. Vaccine **37**(9), 1168–1173 (2019). https://doi.org/10.1016/J.VACCINE.2019.01.033
22. Reiter, P.L., Pennell, M.L., Katz, M.L.: Acceptability of a COVID-19 vaccine among adults in the United States: how many people would get vaccinated? Vaccine **38**(42), 6500–6507 (2020). https://doi.org/10.1016/j.vaccine.2020.08.043
23. Shen, A.K.: Finding a way to address a wicked problem: vaccines, vaccination, and a shared understanding. Hum. Vaccines Immunother. **16**(5), 1030–1033 (2020). https://doi.org/10.1080/21645515.2019.1695458
24. Jarrett, C., Wilson, R., O'Leary, M., Eckersberger, E., Larson, H.J.: Strategies for addressing vaccine hesitancy – a systematic review. Vaccine **33**(34), 4180–4190 (2015). https://doi.org/10.1016/J.VACCINE.2015.04.040

25. Armstrong, K., Ravenell, K.L., McMurphy, S., Putt, M.: Racial/ethnic differences in physician distrust in the United States. Am. J. Public Health. **97**(7), 1283–1289 (2007). https://doi.org/10.2105/AJPH.2005.080762
26. Pilehvari, A., You, W., Marathe, A.V., Anil, S. Black and Hispanic disparity in COVID-19; relative contribution of health care access. Work Pap. (2021)
27. Barkay, N., et al.: Partnership with Facebook (2020). https://dataforgood.fb.com/docs/covid-19-symptom-survey-request-for-data-access/. Accessed 21 July 2021
28. Chen, J., et al.: Prioritizing allocation of COVID-19 vaccines based on social contacts increases vaccination effectiveness. medRxiv (February 2021). https://doi.org/10.1101/2021.02.04.21251012
29. Polack, F.P., Thomas, S.J., Kitchin, N., et al.: Safety and efficacy of the BNT162b2 mRNA Covid-19 vaccine. N. Engl. J. Med. **383**(27), 2603–2615 (2020). https://doi.org/10.1056/NEJMoa2034577

The Spreading-Activation Framework Does not Explain the Effects of Degree and Clustering on Spoken Word Recognition

Leo M. Niehorster-Cook(✉)

Lawrence, USA
LeoNiehorsterCook@gmail.com

Abstract. During spoken word recognition (SWR), various lexical properties affect speed and accuracy, and computational models offer competing explanations for these effects. Classical psycholinguistics has focused on properties internal to individual lexical items such as the facilitative effects of high probability at various levels of processing, rather than focusing on the relations between lexical items. Even connectionist models, which represent SWR as a dynamic process of activation flowing over a hierarchical network of lexical and sublexical representations, have not focused on that network's topology nor on how a word's position in that network affects its recognition.

In contrast, the Cognitive Network Science (CNS) approach has addressed how network topology affects a variety of psycholinguistic tasks. For example, speed and accuracy of perception has been found to be affected by a word's degree on a simple "phonological neighborhood network" (PNN), in which words are connected to similar-sound words. Similar effects have been found for other network properties, and several simulation studies have suggested that (at least some of) these behavioral effects can be explained by a simple spreading-activation algorithm along PNNs, rather than positing a hierarchical network of distributed processing units. In particular, some CNS adherents have claimed that the spreading-activation framework, in combination with the techniques of network science, can explain the effects of degree and clustering on SWR performance.

The present study rigorously explores the state-space of the proposed spreading-activation framework. Adopting the agent-based modeling approach and implementing the process in NetLogo provides a powerful tool to explore this state-space. A series of simulations shows that the framework predicts word-degree in the PNN to have an effect on SWR performance in the opposite direction from observed behavioral results, contrary to previous studies. This suggests previous results are an artifact, presumably resulting from network construction techniques and linear regression modeling that did not reflect the system's complexity. The effects of clustering are also analyzed, which suggest the model's predictions are not as straightforward as has been suggested by prior studies. Implications for models of spoken word recognition are discussed.

The author is not presently associated with any academic institution. The author declares no competing interests. No grants or funding contributed towards the research presented in this document.

Z. Yang and E. von Briesen (Eds.): CSSSA 2021, SPCOM, pp. 112–123, 2022.
https://doi.org/10.1007/978-3-030-96188-6_9

1 Introduction

1.1 Cognitive Network Science & Phonological Neighborhood Networks

Cognitive Network Science (CNS) is a young and burgeoning field. Over the past two decades, there have been many advances shining light on how the structural organization of mental representations influences the processes that play out on these representations. Cognitive network science differs from connectionist approaches to cognition. Connectionist approaches model cognitive *processes* using "artificial neural networks," whose nodes symbolize processing units and whose connections symbolize causal interactions. In contrast, CNS aims to model the *structure* of representational space using "cognitive networks," whose nodes symbolize mental representations and whose links symbolize mathematically- or behaviorally-derived relationships between the nodes such as phonological similarity. For more on the relationship between CNS and connectionism, see Castro and Siew (2020) and Vitevitch and Mullin (forthcoming).

In particular, much CNS research has represented the lexica of human languages using phonological neighborhood networks (PNNs). These networks are made up of nodes representing words. Edges connect similar-sounding words, defined in this context as words whose phonological transcriptions can be converted to each other by the insertion, deletion, or substitution of a single phoneme (i.e., words whose phonological forms are separated by an *edit distance* of 1). For example, the English word "plan" (/plæn/) is connected to "plant" (/plænt/) by addition, "flan" (/flæn/) by addition, and "pan" (/plæn/) by deletion. Pictured here is the local neighborhood (reproduced with permission from Turbull 2021).

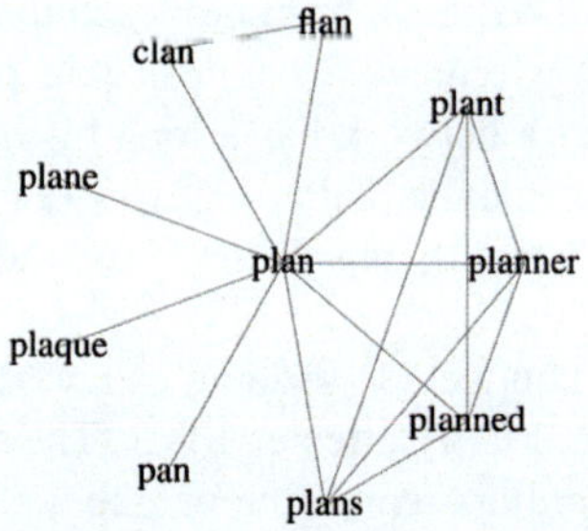

This approach was first applied to English words (Vitevitch 2008) and later extended to other languages (Arbesman et al. 2010). These seminal works compared the PNNs of real languages against randomly generated networks with the same number of nodes and edges. More recent work has compared the PNN of English to PNNs of randomly generated lexica (e.g., Stella and Brede 2015; Turnbull and Peperkamp 2016), and analytic methods are beginning to be applied to exploring the space PNNs may occupy (Turnbull 2021). This line of research seeks to identify characteristic topological properties of PNNs that can inform how we understand the processes that shape human language.

If PNNs robustly display a given topological property relative to randomly generated lexica, that may reflect a selective pressure on language that shapes the lexicon's self-organization. For example, while real lexica yield PNNs with degree distributions best fit by a truncated power law (Arbesman et al. 2010), the PNNs of randomly generated lexica display degree distributions best fit by a (non-truncated) power law, suggesting a systematic tendency for real lexica to avoid including words with an especially high degree (Stella and Brede 2015). Similarly, Turnbull and Peperkamp (2016) identify a robust pattern of high assortative mixing by degree in PNNs of natural lexica. This line of research is young, but it already suggests powerful methods to explore how lexica self-organize.

1.2 Structural Influences on Cognitive Processing

Another line of research has investigated how the lexicon's structure influences performance on spoken word recognition (SWR). When controlling for extraneous variables such as word length, words with high degree are more difficult to perceive than words with few neighbors. The inhibitory effect of a "dense phonological neighborhood" is well-known in psycholinguistics and substantially predates the cognitive network science approach, and competing models of SWR provide alternative explanations of this phenomena. More recent experimental work has explored how topological properties other than a word's degree affect processing, including the clustering in the word's local neighborhood (Chan and Vitevitch 2009; 2010; Vitevitch et al. 2012; Redmon et al. under review) and two-hop neighborhood (Siew 2017), its presence in the giant component of the lexicon vs. in smaller components (Vitevitch and Siew 2016), its "keyplayer status" (Vitevitch and Goldstein 2014), and its closeness centrality (Goldstein and Vitevitch 2017). These findings guide research into language disorders and inform clinical treatments (Vitevitch and Castro 2015; Castro et al. 2017). For a review on the intersection of CNS and psycholinguistics, see Vitevitch (2021).

Computational models of SWR have been proposed to account for these effects. In particular, a spreading-activation framework has been quite popular in the CNS literature. The concept of activation over a network has a long history in cognitive science, but its early use focused on how it may explain *processes* (e.g., Collins and Loftus 1975) and ignored the network structure. Its use in the CNS tradition approach began more recently.

Chan and Vitevitch (2009) conducted a series of behavioral experiments that explored how the clustering coefficient, or C(x), affected SWR. They found that words with high C(x), being words whose neighbors tend to be neighbors with one another, were less easily perceived than words with low C(x). Moreover, they showed that prevailing models of SWR could not explain this effect. Thus, they proposed a novel "spreading-activation framework."

1.3 The Spreading-Activation Framework

In the spreading-activation framework, activation spreads along a network of lexical representations. This model could explain the effects of the PNN's structure on SWR, including the effects of C(x). In particular, in the case of words with high C(x), activation would preferentially flow from the stimulus word's neighbors to other neighbors of the stimulus word, getting "trapped" in the local neighborhood. In the case of words with low C(x), activation would flow from the stimulus word to other parts of the lexicon, more quickly dispersing. This would allow the stimulus word to more easily "stand out" next to its competitor words.

Initially proposed verbally, a computational model subsequently replicated this effect on all the stimulus items used in Chan and Vitevitch (2009) for five time-steps (Vitevitch et al. 2011). Those stimulus items were selected to control for other variables, including the degree of the target node: both the high- and low-clustering word sets had a wide distribution of degree, to ensure that the effects of C(x) generalized and were not confined to, for example, nodes with a small degree. Moreover, the results were analyzed using a linear multiple regression that included three topological regressors - clustering coefficient, degree, and network density - along with one parametric regressor, the retention parameter.

According to their regression, the results replicated the inhibitory effect of high C(x) observed in behavioral experiments (Chan and Vitevitch 2009). That is, words with high C(x) had lower activation values than words with low C(x), indicating that they would be more difficult to perceive. Moreover, they also found that degree had an inhibitory effect on activation - words with a higher degree had lower activation values. This, too, agrees with the effect observed in behavioral data (Vitevitch 2008).

This latter finding was especially surprising and was not the goal of the simulation study. Connectionist models of SWR claim that high-degree words are harder to perceive for one of two reasons - either directly because activated nodes directly inhibit their competitors (e.g., TRACE; McClelland and Elman 1986) or because the choice rule governing selection is sensitive to the activation distributed throughout the system, not just in the target node (e.g., PARSYN; Luce, Goldinger et al. 2000).

The spreading-activation algorithm was also implemented in the R package *spreadr* and used to account for the effects of C(x) on memory tasks and the effects of semantic priming (Siew 2019). The spreadr package has since been used to explore other ways the PNN's structure appears to influence SWR, such as the effects of presence in the giant component and keyplayer status (Vitevitch and Mullin forthcoming). Other research has used the spreading-activation algorithm in combination with other sorts of models of cognitive networks to investigate how their structure may affect processing, including the effects of phonotactic probability on SWR (Vitevitch 2021), the effects of semantic diversity on lexical acquisition (Mak et al. 2021), and the complex facilitatory effects of anodal transcranial direct current stimulation on verbal recall (Díez et al. 2019).

However, it is conceivable that the effects of degree and C(x) on activation value observed in the original implementation are artifacts of the particular topology of the networks under question or due to arbitrary measurement choice. Specifically, Vitevitch et al. (2011) used the two-hop networks of the stimulus words used in Chan and Vitevitch (2009), and measured only at one time-step, $t = 5$. Human languages have complex lexica, which may not well have been accounted for by the modeling techniques used. To explore this possibility and interrogate the spreading-activation algorithm's behavior more generally, the same spreading-activation algorithm was implemented in NetLogo (Wilenski 1999). NetLogo was chosen for two reasons.

First, NetLogo is both a programming language and its own Integrated Development Environment, so the model can be manipulated by the user without the arduous process of learning the language's syntax. The program's graphical user interface is highly accessible, especially for linguists and psychologists, many of whom are not trained in any programming language. NetLogo automatically visualizes the model's behavior in real-time.

Secondly, NetLogo is associated with the agent-based modeling framework. Agent-based modeling is common in the social sciences and is increasing in popularity in the biological sciences, but to date remains relatively unrecognized in the cognitive sciences. Moreover, the agent-based modelling community is heavily intertwined with the network science community, and it also includes several researchers who are interested in the dynamics of diffusion processes in their own right. Introducing the agent-based modeling framework to CNS will propel scientific advances, both due to the influence of agent-based modelers on cognitive science and due to the influence of cognitive sciences on other agent-based models.

2 Simulation 0

2.1 Purpose

Simulation 0 sought to confirm that the NetLogo implementation faithfully replicated the algorithm first described in Vitevitch et al. (2011) and later implemented in the R package *spreadr* (Siew 2019). Since the spreading-activation algorithm is deterministic, only one run is needed on a given parameter-set to confirm faithful implementation. However, various programming languages handle floating-point arithmetic differently, producing slight differences in the results. Further, it is well-known that in complex systems, even minute differences at one point can cause quite large differences to emerge later - the so-called "butterfly effect."

This was unexpected, since (unlike connectionism's artificial neural networks) the spreading-activation algorithm involves no nonlinearity. Still, rather than rely on verbal reasoning and intuition, this was confirmed by simulated spreading-activation using the same parameters and network topologies on both the NetLogo and the R implementations. Given the same input, the resulting output was expected to be the same from both the Netlogo and R implementations.

2.2 Setup

All simulations used regular one-dimensional lattices with twenty nodes. These networks varied in their connectivity; nodes were connected to the nearest 2, 3, or 4 other nodes in each direction, yielding three topologies. Retention also varied across ten levels - from 0.0 to 0.9 by increments of .1. The two implementations were compared across 30 total conditions.

2.3 Results

The activation of the target node was measured at every tick until convergence. Across all conditions, results confirmed that the spreading-activation algorithm was identical. In all conditions, small differences emerged between the NetLogo and R implementations.

These differences were extremely small, with extremes at -6.394885e-14 and 5.684342e-14. These values are several orders of magnitude smaller than the effect sizes reported in studies using the spreading-activation algorithm and should be regarded as negligible.

3 Simulation 1

3.1 Purpose

Simulation 1 served to investigate how the target node's degree and C(x) affect its activation value over many ticks. Rather than extract two-hop networks from specific words in the English lexicon, which could introduce complexities that may confound the results, random networks of 100 nodes were used. This allowed several factors such as network size and topology to be controlled.

3.2 Setup

To investigate the early-vs-late effects of degree and C(x) across a variety of conditions, retention was varied from 0.00 to 0.90 in increments of 0.10. Thus, there were ten retention conditions. Random networks were constructed, where each pair of nodes is connected with probability p, and the presence of each edge is independent of the presence of each other edge. For each retention condition, 1000 random networks with $p = 0.10$ were constructed, for 100,000 total networks. Before analyzing the data, all graphs that were disconnected were removed. When using the spreading-activation algorithm on disconnected graphs, there is no way for the activation to reach components that are not connected to the target node. This artificially decreases the network size, limiting the number of fewer nodes for the activation to reach and serving as a potential confound. This resulted in 289 runs being removed, less than 1% of all runs.

The expected degree of any given node on a random network is equal to the number of other nodes in the network times p. Thus, the expected degree of the target node in the network is $99 * 0.10 = 9.9$. Degree is discrete, so the mode degree is expected to be 10, with a mean of 9.9. In such a network, degree follows the binomial distribution, because it is the sum of independent Bernoulli trials. The expected C(x) of a node on a random network is equal to p, regardless of its degree. When controlling for degree, C(x) is also the sum of independent Bernoulli trials, so it also follows a binomial distribution.

3.3 Results

Because both degree and C(x) follow a binomial distribution in random graphs, which closely approximates a discretized normal distribution for large samples, first-order correlation (Analysis 1) and multiple linear regression (Analysis 2) were used to analyze the effects of degree and C(x). Analysis 1 ignored C(x) and reported the first-order correlation (i.e., Pearson's R) between the target node's degree and its activation over all runs over 50 ticks. The first-order correlation for each retention parameter was analyzed separately. The sign of the correlation determines whether the spreading-activation framework predicts that high degree has a facilitatory effect on processing (positive correlation) or an inhibitory effect on processing (negative correlation). As can be seen in

Fig. 1, at every tick after the second and in all retention conditions, the target node's activation is positively correlated with its degree. This correlation grows monotonically with time, except for when retention is especially low, in which case we see a jagged increase.

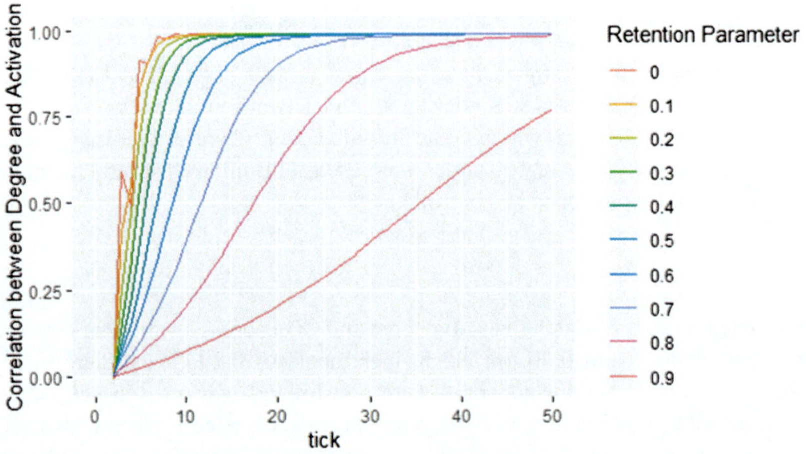

Fig. 1. Correlation between target node degree and activation over time

Analysis 2 considered the effects of degree and C(x) using multiple linear regression. Since the clustering coefficient of a node with degree < 2 is undefined, all networks where the target node did not have degree ≥ 2 were dropped. The total number of edges was also included as a regressor, similar to the multiple linear regression analysis in Vitevitch et al. (2011), which included degree, C(x), and network density. However, that regression model also included the retention parameter as a regressor and predicted the node's activation value only at $t = 5$; Analysis 2 is a set of regression models, one for each combination of retention parameter and time step between $t = 2$ and $t = 200$.

There was no reason to suspect C(x) or edge count was confounding the results of Analysis 1, but to protect against that possibility, the coefficient for degree as well as clustering is reported. Figure 2 shows these results. As can be seen by visual comparison to Fig. 1, Analysis 2 agrees with Analysis 1, confirming that C(x) did not serve as a compound. Degree has a uniformly positive effect on activation, an effect which monotonically grows and is never negative.

Analysis 2 also allowed us to interrogate the effects of C(x) on activation. The sign of the parameter for the clustering coefficient determines whether high C(x) has a facilitatory effect on processing (positive parameter) or an inhibitory effect on processing (negative parameter).

Figure 3 shows the results of Analysis 2 for C(x). These figures display complex results. First, regardless of retention, C(x) has an initially negative effect.[1] Secondly,

[1] Note that, since network structure cannot affect the target node's activation until tick 2, the data displayed here begin on tick 2.

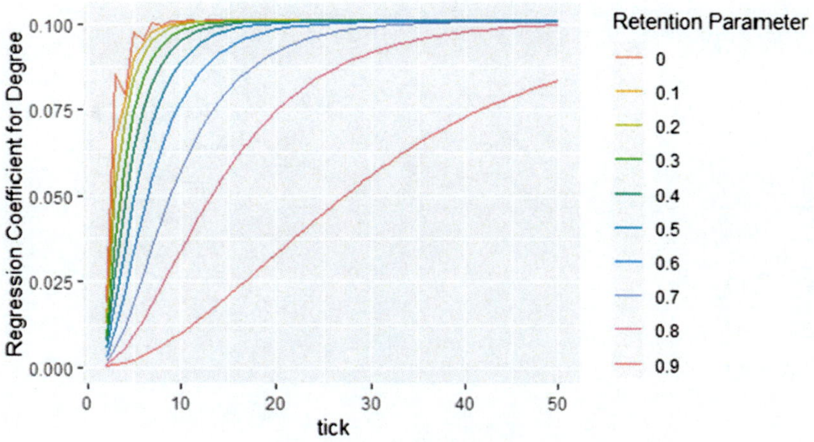

Fig. 2. Analysis 2 - degree parameter over time

when retention is low, the clustering parameter changes signs on each tick towards the beginning of the simulation, positive on odd ticks and negative on even ticks, and diminishing in absolute value. For moderate to high levels of retention, there is an early negative effect which first grows in magnitude, then shrinks and eventually switches signs to a small positive effect, and then eventually back to a small negative effect. These models are converging to a point where this effect disappears entirely.

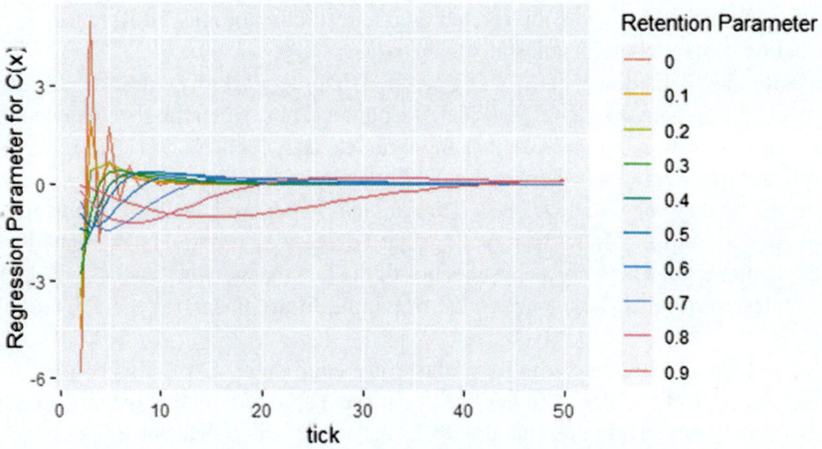

Fig. 3. Analysis 2 - clustering parameter over time

Due to the extreme values taken on by aberrant low retention groups, Fig. 4 shows the same results, but only for runs where the retention parameter was ≥0.5 and for up to 200 ticks.

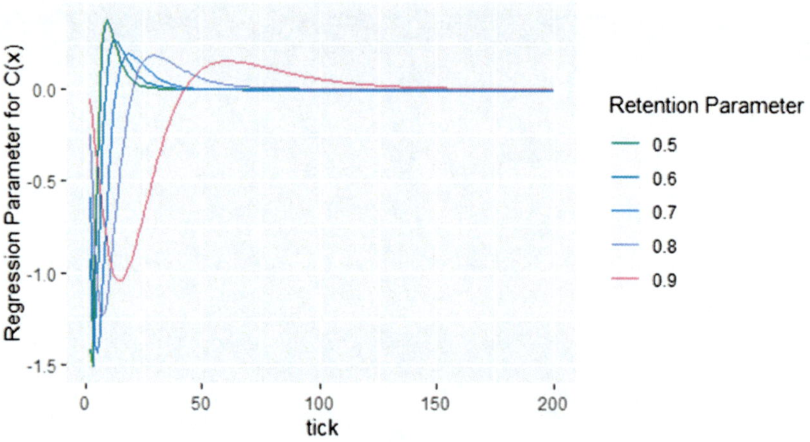

Fig. 4. Retention ≥ 0.5, Analysis 2 - clustering parameter over time

3.4 Discussion

The results of Simulation 1 generally conflict with the results of Vitevitch et al. (2011).

In the case of degree, Vitevitch et al. (2011) found that nodes with a higher degree had lower activation values. This lined up with the behavioral data, suggesting that the spreading-activation framework can explain the effects of degree on SWR. This conflicts with the present results, which show that nodes with a higher degree have higher activation values. Since these data were generated using the same spreading-activation algorithm, this was interpreted as due to methodological differences in the construction of the networks and statistical analysis.

In particular, Vitevitch et al. (2011) generated their networks by extracting the stimulus items' 2-hop networks from the PNN of English. Thus, when the stimulus item has many neighbors, the 2-hop network has more nodes, introducing a correlation between network size and degree. Furthermore, since the PNN of English exhibits high assortative mixing by degree (Vitevitch 2008; Turnbull and Peperkamp 2016), this correlation was likely superlinear. Thus, their finding that degree was negatively correlated with activation after 5 ticks is likely a consequence of the fact that networks with high-degree target nodes were much larger networks, providing more places for the activation to spread.

The present results concerning the clustering coefficient, C(x), also conflict with those of Vitevitch et al. (2011). In general, the present study found that the target node's C(x) is initially negatively correlated with its activation value, but this effect switches direction. Since Vitevitch et al. (2011) only measured at $t = 5$, they found this negative effect. However, there is no theoretical motivation for preferring results at $t = 5$ over other time-points. Moreover, when the retention parameter is ≤ 0.2, the correlation has become positive by the fifth tick. It is thus unclear whether the model should be interpreted as predicting the early inhibitory effect or the late facilitatory effect, and theoretical arguments would be needed to interpret the model as replicating the effect seen in behavioral data.

4 General Discussion and Conclusions

4.1 The Spreading-Activation Algorithm

The simulations reported here identified a robust pattern of high-degree nodes retaining more activation than low-degree nodes. This effect is inconsistent with the behavioral results concerning the PNN's effect on SWR. Analysis 2 shows the effect of C(x) is dependent on when the modeler chooses to measure, with an early effect consistent with the behavioral results and a late effect inconsistent with the behavioral results. This indicates that the model is able to predict either pattern, introducing difficulties into interpreting the data.

However, this does not entail that the spreading-activation algorithm should be abandoned. Future research in cognitive network science may uncover situations in other domains where high-degree nodes are more easily accessed. In these situations, the spreading-activation algorithm provides powerful hypotheses about the underlying nature of that effect – when activation is diffusing through a network, it tends to aggregate at high-degree nodes. Indeed, there is evidence that high degree on the PNN facilitates lexical perception during reading (Yates 2004; Yates et al. 2008). Future research should investigate whether these effects can be attributed to a spreading-activation process.

4.2 Cognitive Network Science and Agent-Based Modeling

More generally, the present study may serve as a first step in the application of the agent-based modeling framework to CNS. As a field, CNS aims to use networks to model the representational space within which a cognitive system must function, and to study how the structure of those networks affects processes that play out over those networks. Controlled laboratory experiments and naturalistic observation techniques reveal that network structure affects cognitive processing, but they do not reveal the underlying mechanisms. Agent-based modeling is well-suited to explore how behavioral effects may emerge from processes playing out over complex networks.

Following a tradition of simulation studies in the psycholinguistics literature, the present study interpreted the agents as lexical representations. Future research in cognitive network science may choose other cognitive constructs to instantiate as agents in agent-based models. These may include high-level representations such as beliefs and attitudes or low-level constructs such as perceptual features. Psychopathological models may interrogate how representational space influences the processes involved in psychological illness. This approach may be in harmony with the network model of mental disorders, which describes psychological illness as a complex interaction between symptoms rather than a set of symptoms caused by some latent disease entity (Borsboom 2017).

This is a departure from the traditional interpretation of agents in computational social science. As one reviewer notes, traditional CSS approaches "[…] prefer to observe an agent as one who speaks or interprets/recognizes spoken words and possesses clinical symptoms and neurons that factor into the human's ability to process spoken words." The CNS approach provides a "scale-invariant" toolkit to study multiple levels of representation. What is consistent is a focus on the agents' organizational structure and how this affects processing.

5 Additional References

All relevant code, including the NetLogo model, the simulation data, and the analysis, is available at https://github.com/LeoMNC/Spreading-Activation/. All statistics were calculated using Base R, version 4.0.5 (R Core Team 2020). Data were manipulated using the tidyverse (Wickham et al. 2019). The ggthemes package was used in figure creation (Arnold and Arnold 2015). Regular rings were created in R for Simulation 0 using igraph (Csardi and Nepusz 2006).

References

Arbesman, S., Strogatz, S.H., Vitevitch, M.S.: The structure of phonological networks across multiple languages. Int. J. Bifurcat. Chaos **20**(03), 679–685 (2010)

Arnold, J.B., Arnold, M.J.B.: Package 'ggthemes' (2015)

Borsboom, D.: A network theory of mental disorders. World Psychiatry **16**(1), 5–13 (2017)

Castro, N., Pelczarski, K.M., Vitevitch, M.S.: Using network science measures to predict the lexical decision performance of adults who stutter. J. Speech Lang. Hear. Res. **60**(7), 1911–1918 (2017)

Castro, N., Siew, C.S.: Contributions of modern network science to the cognitive sciences: revisiting research spirals of representation and process. Proc. R. Soc. A **476**(2238), 20190825 (2020)

Collins, A.M., Loftus, E.F.: A spreading-activation theory of semantic processing. Psychol. Rev. **82**(6), 407 (1975)

Chan, K.Y., Vitevitch, M.S.: The influence of the phonological neighborhood clustering coefficient on spoken word recognition. J. Exp. Psychol. Hum. Percept. Perform. **35**(6), 1934 (2009)

Chan, K.Y., Vitevitch, M.S.: Network structure influences speech production. Cogn. Sci. **34**(4), 685–697 (2010)

Csardi, G., Nepusz, T.: The igraph software package for complex network research. InterJournal Complex Syst. **1695**(5), 1–9 (2006)

De Deyne, S., Perfors, A., Navarro, D.J.: Predicting human similarity judgments with distributional models: the value of word associations. In: Proceedings of COLING 2016, the 26th International Conference on Computational Linguistics: Technical Papers, pp. 1861–1870 (2016)

Díez, E., Alonso, M.A., Díez-Álamo, A.M., Gómez-Ariza, C., Fernandez, A.: False recognition reduction effects induced by anodal tDCS and its relation to different measures of expected spreading activation for critical words (2019)

Luce, P.A., Goldinger, S.D., Auer, E.T., Vitevitch, M.S.: Phonetic priming, neighborhood activation, and PARSYN. Percept. Psychophys. **62**(3), 615–625 (2000). https://doi.org/10.3758/BF03212113

Mak, M.H., Hsiao, Y., Nation, K.: Anchoring and contextual variation in the early stages of incidental word learning during reading. J. Mem. Lang. **118**, 104203 (2021)

McClelland, J.L., Elman, J.L.: The TRACE model of speech perception. Cogn. Psychol. **18**(1), 1–86 (1986)

R Core Team: R: A language and environment for statistical computing. R Foundation for Statistical Computing, Vienna, Austria (2020). https://www.R-project.org/

Siew, C.S.Q.: spreadr: an R package to simulate spreading activation in a network. Behav. Res. Methods **51**(2), 910–929 (2019). https://doi.org/10.3758/s13428-018-1186-5

Stella, M., Brede, M.: Patterns in the English language: phonological networks, percolation and assembly models. J. Stat. Mech: Theory Exp. **2015**(5), P05006 (2015)

Turnbull, R.: Graph-theoretic properties of the class of phonological neighbourhood networks. In: Proceedings of the Workshop on Cognitive Modeling and Computational Linguistics, pp. 233–240 (2021)

Turnbull, R., Peperkamp, S.: What governs a language's lexicon? Determining the organizing principles of phonological neighbourhood networks. In: Cherifi, H., Gaito, S., Quattrociocchi, W., Sala, A. (eds.) Complex Networks & Their Applications V. COMPLEX NETWORKS 2016. SCI, vol. 693, pp. 83–94. Springer, Cham (2016). https://doi.org/10.1007/978-3-319-50901-3_7

Vitevitch, M.S.: What can graph theory tell us about word learning and lexical retrieval? J. Speech Lang. Hear. Res. JSLHR **51**(2), 408–422 (2008). https://doi.org/10.1044/1092-4388(2008/030)

Vitevitch, M.S.: What can network science tell us about phonology and language processing? Top. Cogn. Sci. **14**, 127–142 (2021)

Vitevitch, M.S., Castro, N.: Using network science in the language sciences and clinic. Int. J. Speech Lang. Pathol. **17**(1), 13–25 (2015)

Vitevitch, M.S., Chan, K.Y., Roodenrys, S.: Complex network structure influences processing in long-term and short-term memory. J. Mem. Lang. **67**(1), 30–44 (2012)

Vitevitch, M.S., Ercal, G., Adagarla, B.: Simulating retrieval from a highly clustered network: Implications for spoken word recognition. Front. Psychol. **2**, 369 (2011)

Redmon, C., Tremblay, A., Vitevitch, M.S.: Tracking the time course of phonological neighborhood clustering effects in spoken word recognition (under review)

Siew, C.S., Vitevitch, M.S.: Spoken word recognition and serial recall of words from components in the phonological network. J. Exp. Psychol. Learn. Mem. Cogn. **42**(3), 394 (2016)

Wickham, H., et al. (2019)

Wilensky, U.: NetLogo. Center for Connected Learning and Computer-Based Modeling, Northwestern University, Evanston, IL (1999). http://ccl.northwestern.edu/netlogo/

Yates, M., Locker, L., Simpson, G.B.: The influence of phonological neighborhood on visual word perception. Psychon. Bull. Rev. **11**(3), 452–457 (2004). https://doi.org/10.3758/BF03196594

Yates, M., Friend, J., Ploetz, D.M.: The effect of phonological neighborhood density on eye movements during reading. Cognition **107**(2), 685–692 (2008)

Engineering Decentralized Enterprises: Emergent Mission Accomplishment Without Centralized Command and Control

Michael D. Norman(✉), Paul E. Silvey, Matthew T. K. Koehler, and Kirbi C. Joe

The MITRE Corporation, 202 Burlington Road, Bedford, MA 01730, USA
mike.d.norman@gmail.com

Abstract. Rigorous approaches to the engineering of decentralized, multi-agent systems are nascent. Providing contrast to these are hierarchical systems, in which central operational entities distribute tasks and aggregate results. We explore self-organized networks of sensors and effectors to understand how systems without centralized command and control can be created to accomplish an enterprise-wide mission when the individual agents do not have a concept of command or hierarchy. The individual agents do, however, have knowledge of how to act given their own embodied situational awareness, with an implicit understanding that these behaviors contribute to collective enterprise mission goals. When such local situation awareness is shared with appropriate other agents, we say the overall system is using *distributed situational awareness (DSA)*. Our practical systems engineering challenges then are to determine what kinds, how much, and through what mechanisms such DSA should be shared, to enable the overall system to be most effective and efficient. The problem used to explore these questions is one of mitigating the damage caused by asteroid impacts within the continental United States with a set of autonomous vehicles that have limited capacities for sensing, communications, and information processing. We use this challenge problem to prototype an enterprise command and control system using an approach rooted in complexity science and computational social science methods. The experimental testbed we describe here allows us to study emergent engineering of decentralized multi-agent systems, by observing their enterprise-level mission performance as we apply stressors and instill agent-level behaviors enabled by DSA information sharing. Interestingly, we find that maximum system-level performance is obtained when 1) the vehicles are able to minimally coordinate mitigation efforts, and 2) are of limited capability with respect to sensor and communications ranges.

Z. Yang and E. von Briesen (Eds.): CSSSA 2021, SPCOM, pp. 124–151, 2022.
https://doi.org/10.1007/978-3-030-96188-6_10

1 Introduction

1.1 [Centralized] Command and Control

The concept of Command and Control (C2) emerged in the context of military decision making but is generally applicable to multi-agent collective problem-solving in dynamic, uncertain environments (often adversarial or competitive ones). Although many C2 problems are modeled or studied as games with strict constraints such as rules of engagement and bounded numbers and capabilities of agent participants, the real world is much more open-ended. To operate with a singularity of purpose despite a massive distribution of labor, hierarchical organizations emerge, centralizing their strategic decision-making to command levels where the most difficult cognitive processes are concentrated. However, if all relevant information for choosing a course of action were known at the level in which material actions occur, the centralized approach might be replaced by a fully decentralized one. The difficulty of achieving this lies in the fact that neither the necessary information per node, nor the action policies based on that information are easily determined to accomplish enterprise mission objectives under a wide range of situational eventualities.

As noted, command and control of enterprise decision-making has historically been a highly centralized activity. Commanders make decisions on desirable enterprise-scale outcomes, and those desired outcomes are then pushed down through the C2 echelons and successively decomposed into subplans based on the decisions of those at each echelon. Execution of a commander's orders typically follows a hierarchy of C2 by way of echelon. This form of regimented C2 is the current doctrine.

The concept of situational awareness (SA) is used in centralized enterprises to refer to information about the system and its environment that is available to the central command node to inform decision-makers. The SA available to a central authority node is often limited in scope, lossy and potentially stale by the time it reaches decision-makers. In centralized C2, SA information filters up, but it rarely is pushed back down, creating a tension within the tactical components who are needed to collect information about the environment but who rarely benefit from the effort.

1.2 Decentralized Command and Control

Decentralized command and control can only be realized if nodes are provided a significant level of agency and the command hierarchy is flattened. The term agent in this context refers to nodes that have been imbued with the ability to participate in adaptive decision-making, inclusive of leaf nodes at the tactical edge. The scope of their decision-making is concentrated on their local physical actions including (but not limited to) their mobility, but it also includes choices they make with respect to information processing and communication. The information used to inform their decisions is gleaned from local observations of their environment and information shared with them by other agents residing in the communications network. No single agent is expected to have complete situational awareness of the entire system, nor should we expect any two agents to have synchronized, or even consistent, observations or perceptions of its state. Although more complex than a highly optimized, functionally segregated enterprise, enabling agency may lead to a more resilient and less fragile system as it empowers individual functional units to deal with uncertainty in the proximity of its occurrence.

Information Loss. In a centrally controlled system, data is lost when an edge node must compress its local situational awareness down to a transmittable and digestible chunk for consumption by a central authority [1]. Although this loss of data is often by design, especially in contested environments, the potential information/knowledge/wisdom that may be generated from this data is limited by the decisions of the intervening echelons between the commander and the leaf node. A node may face a manifestation of uncertainty in this process resulting in loss of data that previously held no value under conditions of certainty. An additional challenge is that neither the central authority nor the communications network can feasibly handle the raw experiential data being produced by all the nodes within an area of responsibility. The data lost may well contain vital clues as to what the optimal decision should be in the current situation. This may result in potentially bad action decisions being pushed out to the leaf nodes. The information loss inherent to a hierarchical system architecture created to support centralized decision-making may be presented as an unavoidable technological challenge to be overcome, rather than an inherent effect of system design decisions. This is often seen when an enterprise seeks to collect all data under one roof. However, even if all enterprise data sources are federated, and not aggregated, the meta-strategy of "aggregate all the data first" misses the true challenge: even if one has "all the data", one does not have the path-dependent experience of the embodied leaf nodes that collected that data (nor the time to manually review details at that scale). Information with premeditated value is generated for a disembodied authority which is subsequently challenged by uncertainty when it presents itself in the realm of the data the system was designed to propagate upwards. "All the data" is never all the data; it is the data that was preconceived to have value under conditions of certainty.

Timeliness. Timeliness becomes a challenge for centralized C2 as there is an inherent delay, especially in contested environments, between an edge node encountering an unpredicted event, generating a report of that event (which is subject to the information loss discussed in the previous section), transmitting that report, and awaiting the proceeding command. Like the information loss issue, often the timeliness issue is lensed as a technological obstacle to be overcome, rather than an unavoidable consequence of centralized system design.

In a centralized design, the likelihood of the inability to propagate information from any node to the centralized node is inherently less likely than if there were many possible decision-making nodes. This is of course a consequence of the topology. All paths that must route to a particular node have a lower probability than if there were either many pathways to that node, or many alternative nodes. The longer the pathway the more likely an error and, likely, increasing delivery time.

Can we break out of this paradigm? Can we design an enterprise that can accomplish its mission without centralized decision-making? The simulation formulated to explore our challenge problem is abstract and coarse-grained to push forward in answering these questions while embracing the holistic complexity of the entire system of agents operating simultaneously.

1.3 Mission as an Emergent System Property

When a decentralized system is operating, its performance is emergent, as the system level dynamics are not reducible to any individual agent. If the performance is in service of achieving a commander's intended outcome, then we may view the mission itself as an emergent property of the system. This is especially true because mission objectives are usually achievable only to a degree, and the outputs of complex adaptive systems are frequently sensitive to small changes in initial conditions or configuration parameters.

Although the general understanding is that there are certain mission-scale decisions that must be made by a centralized authority, the underlying assumption for our current research is that decisions occurring within the execution of many enterprise-supporting functions can be delegated downwards. Future work will include identification and modeling of the decisions of an enterprise that cannot be decentralized. We believe that properly engineered adaptive and self-organized enterprises should be capable of operating in decentralized ways most of the time yet be able to respond to truly unexpected situations by appealing to some forms of centralized command and control. To that end, we begin by exploring architectures where all decisions are localized and distributed, deferring command authorities to future situations where centralization is strongly justified or needed.

1.4 Distributed Situational Awareness

Distributed situational awareness (DSA) is the information that any agent in the system has about the system itself, its environment, and information collected by interacting with its network peers. Any two agents do not likely retain the same perception of state due to observational variations and inherent truth distortions (e.g., two agents may have temperature sensors, but they are not exactly calibrated). We consider the set of data flowing through the system specific to the system-level goals to be "mission information" (MI)—this is typically the only information that a non-command node in a centralized enterprise would have access to. The set of data that exists in a decentralized enterprise's agents which is not in the set of MI, whether it is information an agent has about itself, its peers, or its environment (whether obtained via sensing, processing, inference, or other means), and here is referred to as DSA. The experimental testbed we describe here allows us to study emergent engineering of decentralized multi-agent systems [8,13,27,33], by observing their enterprise-level mission performance as we apply stressors and instill agent-level behaviors enabled by DSA information sharing, discussed *infra*.

1.5 The Challenge Problem (Asteroid Impact)

To explore how a mission can be accomplished by a decentralized system, a challenge problem was created. The challenge problem is an autonomous asteroid impact response system (AAIRS). The stylized simulation is composed of a population of autonomous vehicles (AVs), each of which has an ability to communicate, and some of which have specialized capabilities for sensing nearby asteroid impacts and/or effectors which can repair the harm caused to people and critical infrastructure by nearby asteroid hits. The AVs are spatially confined to the continental US (CONUS). Asteroid impacts resulting from the Earth's passage through the tail of a comet induce perturbations to the system that affect its performance. The system must locate (sense) these impacts, dispatch repair (effector) units, and execute repair operations as quickly and with as little agent attrition as possible by interacting with one another within a shared environment and through their communication links.

2 Foundational Prior Work

2.1 Cooperative Distributed Problem Solving vs. C2

There is a rich literature in group decision making, organizational structure and learning, and distributed multi-agent problem solving, both from ethnographic studies of humans in teams to distributed artificial intelligence in multi-agent simulations [1,5,9,28]. This research shows that shared situation awareness is an unrealistic objective in practice, being better characterized as compatible situation awareness that both allows for and leverages individual differences in experience and local perspective, where cognition is seen as socially distributed. In Command and Control (C2) contexts this has resulted in evolution of requirements from seeking a Common Operational Picture (COP) to supporting a User-Defined Operational Picture (UDOP) [16,34]. Unfortunately, because there still exists a desire to have common displays for decision makers showing a God's Eye View of the problem environment, even UDOP implementations typically customize only via filtering by data type. If such filtering is done at display time, the overall system still assumes information relevant for any participant will be available at any location, which is both impractical and unachievable at scale.

In response, it may be argued that centralized C2 approaches try to concentrate knowledge within the enterprise to drive uncertainties out of the decision-making process. The irony of this approach is that the further that the strategic or tactical mission decomposition is taken, the more fragile the system becomes to uncertainty and unpredicted events that are virtually guaranteed to occur [6,30].

2.2 Previous Work on the Study of Complex Systems

There is a large body of work around the modeling and simulation of decentralized systems. Much of this work exists under the label of 'complex adaptive systems' [15]. Although much of the research which has leveraged agent-based modeling tools has been focused on capturing the decentralized interaction mechanisms found in the complex adaptive systems that have evolved in nature [25,26], we adopt the techniques pioneered by this field to inform and guide our prototypical decentralized enterprise engineering process. Whether an agent represents a bird or an unmanned aerial vehicle, the abstract and coarse-grained nature of the holistic problem space around emergent functionality is identical. For example, public blockchain functionality is as emergent as that of an ant colony [20,36].

There has been some work to combine the study of decentralized C2 with Agent-Based-Modeling, and we see our work here as continuing this tradition [4,7,14,29,37]. The trade space is enormous: do we want many, inexpensive agents, or fewer, exquisite agents [12]? Do we want homogenous agents [26] or heterogenous agents [23]? Do we want segregated sub-networks or a single global interconnected network? Do we want to imbue agents with heuristic-based local decision-making rules, or do we want to use machine learning to train them to produce opaque but performant reinforcement learning policies [21]? The only way to answer these questions is to form a hypothesis and then create a simulation to test the hypothesis. Since emergent system behavior is definitively irreducible and therefore does not lend itself to closed-form analysis, we must create an agent-based simulation to generate data which can then be analyzed to determine performance [22].

3 Simulation Approach

3.1 Why Use an Agent-Based Model?

How does one begin to approach the challenging task of creating a decentralized system? Since the functioning of a decentralized system is emergent [19], it can only occur if the entire system is executing. This stands in stark contrast to traditional systems engineering approaches which tend to develop unique functional units in isolation, then permanently compose those functional units to assemble the enterprise workflow. In traditional systems engineering, the whole is equal to the sum of the parts. To engineer decentralized enterprises that accomplish their missions by leveraging emergence, where the whole is greater than the sum of the parts, a coarse-grained and holistic starting point is needed. Agent-based modeling tools are purpose-built to address this decentralized system design challenge [11,17,22,25] because they provide the primitives needed to rapidly prototype coarse-grained systems composed of individual decision-making entities as unified, executable wholes.

In this initial step of breaking out of the traditional systems engineering paradigm to one that can explore decentralized systems, some simplifying assumptions about the enterprise must be made. Firstly, the enterprise in this prototype has but one mission, and that mission does not change. In reality, and in further iterations of this work, it is imperative that the enterprise be capable of adapting itself on a fundamental level, i.e., not just enable agent-scale decisions from moment to moment (which are the subject of this prototypical effort and of most complex adaptive systems research), but also adapting the set of action(s) that an agent is afforded, the heuristics and/or policies that the agent follows when determining whether or not to take said action(s), and how/when they interact with their peers, etc.

Secondly, to enable each agent in our simulation to make appropriate local decisions at run-time, we will need to consider what actions they can take and what information might best inform them in choosing one action over another. Therefore, we see ourselves as system engineers as making design-time decisions about the types of information and action decisions we will give our agents. Then, through extensive simulation experimentation, we collect system level performance metrics to assess how well our design choices work for the enterprise under a wide variety of parameter settings that alter the enterprise configuration. In following this basic methodology, we have found it useful to be guided by the following principles learned from complexity science.

3.2 Four Pillars of Functional Emergence

The study of complex adaptive systems that have evolved in nature provide us with a starting point to begin to assemble and implement some of the concepts that will be necessary for emergent mission accomplishment. We believe these concepts are decentralized decision-making, diversity, functional degeneracy, and stigmergy.

Decentralized Decision-Making. The worst outcome for an agent making a bad local decision is that they may be attritted. In a decentralized system of many agents, this won't have much, if any, effect on the system's continued functioning. On the other hand, in a centralized enterprise, a bad decision by a central authority could cause systemic ruin.

Diversity. Homogeneity across a decentralized enterprise leads to systemic fragility and/or prohibitive cost. Certain agents require certain capabilities, and if every agent has every capability, we will end up with extremely expensive assets. This inflated cost would reduce the number available, and we would end up with a system that is fragile to attrition due to the relatively low number of agents. Additionally, homogenous agents are more susceptible to common-mode failures. This, too, leads to systemic fragility to intelligent perturbation.

According to Ashby's Law of Requisite Variety [2], the complexity of a given environment must be at most equal to the complexity of the interfacing system at the scale of that interface point. If we can design a system with the diversity necessary to generate the complexity required at the scale of the interface point with the environment (i.e., a single agent), we have a system that may thrive. The potential additional benefits of diversity go far beyond the scope of this section and this paper [24].

Functional Degeneracy. A decentralized system must maintain functional redundancy despite diversity. Functional redundancy is typically achieved in enterprise architectures via standby nodes in fail-over roles. Taking our cues from nature's laboratory, we notice that one of the most complex systems in the world, the human brain, doesn't have a backup brain on standby. Instead, functional redundancy to ischemic events is achieved by the plasticity of alternative, self-organized neuronal ensembles [10]. Put simply, degeneracy is when a system's function (or a satisficed version of it) can be performed by diverse agents (or neurons, if you like). An agent or network of agents which are specialized for one task may be able to take on a task meant for other agents which are specialized for another task (although the effectiveness may be adversely impacted; it is far superior to system failure.

Stigmergy. If we are to create a decentralized system, then we must be as elegant and minimalist as possible with communications load to ensure agents with heterogenous comms and processing abilities can enable degeneracy within the system. We are trading off heavy amounts of communication and situational awareness aggregation for increased decision-making at the edge. If decisions are being made at the edge, then direct communication of SA that may be observationally apparent to local agents may be unnecessary.

Stigmergy is traditionally known in the sociobiological literature as a class of mechanisms that facilitate interaction between animals [31]. One of the major features of stigmergy is its focus on indirect communication via the perturbation of a shared environment. Another way to look at it is as a set of instinctive inferences that an animal makes about the state and dynamic activity of other animals of relative import. The coordination and subsequent emergent behavior of an ant colony is enabled by the stigmergic effects of ant pheromones [36]. We consider stigmergy to be a type of DSA. It is important to stress that while information supplied to an agent in system via stigmergy can be transformed into a message which can then be directly communicated (i.e., to a centralized decision-making agent), that information is now taking up bandwidth in the communications network, and compute/storage at each node it must then traverse.

Additionally, direct communication doesn't have environmental persistence the way stigmergic communication may. During direct communication, the system, rather than the environment, is responsible for persisting the message. In decentralized environments, this may prove to be onerous at scale if not approached from a minimalist perspective.

3.3 Optimizing to a Scenario

One of the hypothesized factors that will drive system robustness is not optimizing to a specific premeditated scenario. To avoid this, our baseline threat model, i.e., pattern of asteroid impacts, is stochastic in nature. By building the first iteration of our system to be robust to randomized perturbation, we hypothesize we may develop a design solution that gives us more generalizable results.

4 Simulation Implementation

The simulation we have developed to explore this problem space is called the Autonomous Asteroid Impact Response System (AAIRS) and is implemented in the NetLogo programming environment [32,35]. The environment is a stylized, two-dimensional representation of the continental United States (CONUS). The overall enterprise mission objective, as noted above, is to use agents with sensing capabilities to detect asteroid impacts and agents with effecting capabilities to go to those impact sites and do repair work, and to do all this as quickly and thoroughly as possible while minimizing losses. The essential Mission Information comes from sensing agents when they generate time and location reports of asteroid impacts, as these observations drive overall system response behaviors. Other types of information observations and exchanges constitute our DSA extensions. Before we describe the decentralized approach implemented in our experiments, let's consider how this problem might be solved using traditional C2 hierarchies.

A centralized design might utilize predefined tasking to position the sensor and effector network and its communications support relays into Areas of Responsibility (AORs), assuming that there was enough time to prepare for the anticipated threat. It would then direct the network to funnel all asteroid impact reports to a centralized location, and using such global knowledge, do resource allocation and planning for effector work efforts. Finally, it would send command messages to specific effectors to direct them to their assigned tasking. A central C2 authority would also have to respond to problems introduced because of node attrition, the first being simply to detect losses after they occur, and then to use dynamic planning to appropriately re-assign or backfill for those lost nodes, be they sensors, comms relay nodes, or effectors. And in a worst-case scenario, the enterprise must have contingency plans for what to do if the centralized C2 authority itself is lost.

In the proposed decentralized approach, several aspects of the simulation and how the agents work to solve the challenge problem are driven by random processes, including giving them random starting positions and default random movement behaviors. In addition, the spatial and temporal pattern of asteroid impacts is determined by a stochastic process, by default simply a small but uniform random probability of an impact occurring on the next simulation time tick and if occurring, an equiprobable chance of hitting any CONUS map location.

In addition to these randomized elements, there are several key aspects of the decentralized solution that leverage agent memory and self-directed autonomy. The first is simply that the sensors are created with their sensing capabilities turned on, such that they automatically report the time and location of observed asteroid impacts, as Mission Information (MI) to be shared with whatever nearby agents can hear their communication broadcast. Because of limited signal range, the agents therefore rely on the dynamic network that emerges from unplanned encounters between agents. Sensors and Effectors have additional specialized behaviors, but all agents participate in the comms network using the default autonomous communication behavior. This manifests itself as each agent immediately broadcasting an MI report (starting with generated sensor observation reports), provided they have not already relayed that particular message on an earlier time step. This constitutes a basic message flooding protocol over the emergent network, which only guarantees message propagation within components of the network that are fortuitously connected to the original sensor within the time it takes for the message to transit the number of hops in the diameter of the subgraph.

The effector agents have memory to keep track of asteroids they have learned about and decentralized decision-making heuristics encoded as algorithms to decide which one they should dispatch themselves to. When they complete repair work at a mission site, they either select another target from their list, or the revert to the default random walking behavior they started with.

Additional simulation details can be found in Appendix A: ODD+D Protocol [17]. A final point regarding validation of the simulation, at this point the simulation is being used as a thought experiment and is of a system that does not currently exist (thus, we have no real-world comparator); therefore, we have opted for simple Empirical Relevance Level 0 [3] and simple face validation. This being the case, we only insist that the simulation not behave implausibly, rather than it behaving in a particular way in relation to another referent system.

The methodology for the simulation's implementation can be found in Appendix C.

4.1 Experimental Simulation Parameters

Figure 1 shows an image of the basic simulation. There are three types of agents: Sensors, which sense asteroid impacts; Effectors, that mitigate asteroid damage; and Communication nodes, which relay messages. During the simulation asteroids impact the continental US. The agents are tasks with 1) finding the impacts, 2) communicating the locations to each other, and 3) mitigating the damage caused by the asteroid impacts. Additional details can be found in Appendix A.

4.2 Experiment Execution

Simulations were executed using NetLogo's BehaviorSpace module, which facilitates the initialization of agents within a specified environment, allowing NetLogo to explore emergent behaviors of these kinds of multi-agent systems, and to log their results for subsequent analysis. NetLogo was also used to compute the necessary measures needed to evaluate model performance. The simulation parameters we used with their combinatoric value-swept settings are given in Appendix B.

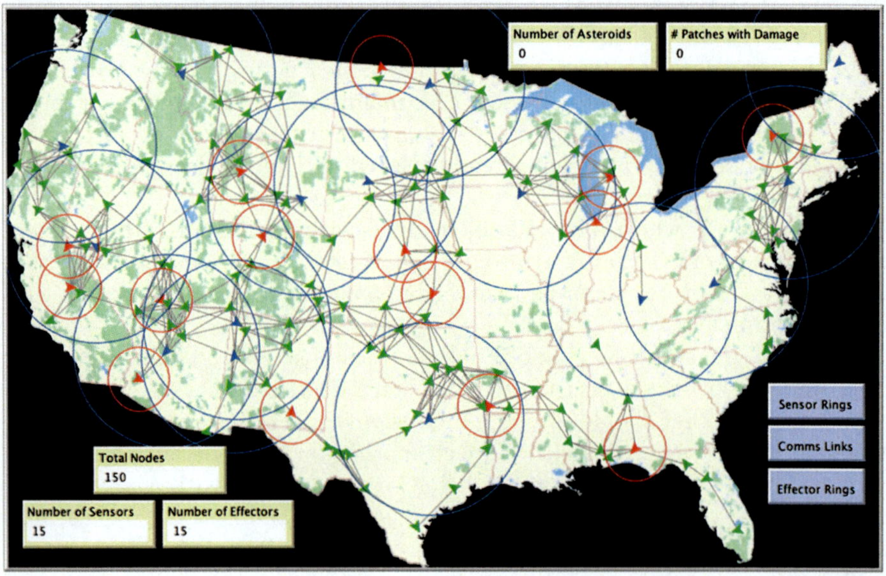

Fig. 1. Example simulation initial conditions

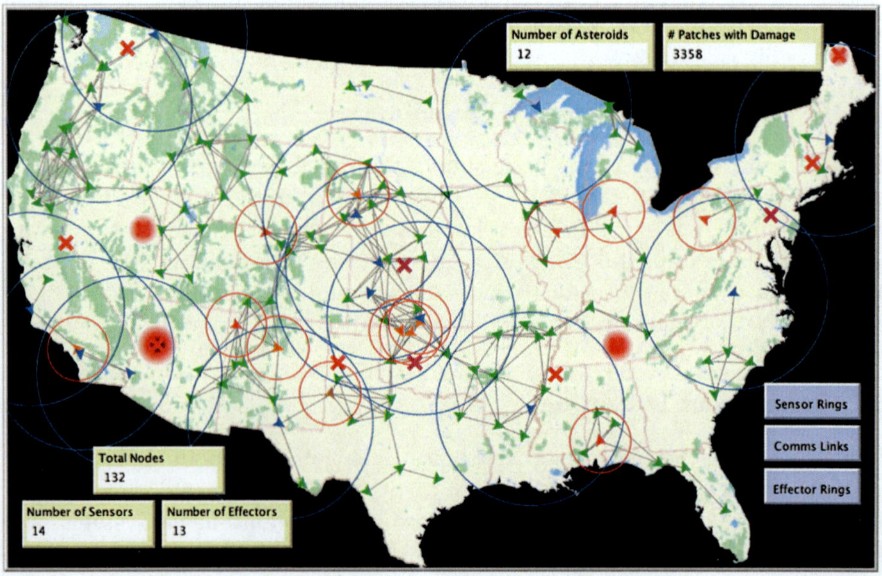

Fig. 2. Example simulation in progress

5 Design of Experiments and Results

5.1 Parametric Changes

Two phases of experimentation were conducted with the simulated system. The first experiment focused on evaluating the performance of the DSA-enabling parameters. The second experiment focused on evaluating the performance of systems composed of fewer exquisite assets against systems composed of numerous cheaper assets via nine different DSA-enabled system specifications.

5.2 Experiment 1: Effects of Toggling DSA Parameters

The goal of Experiment 1 is to establish the validity of the hypothesized DSA parameters by comparing the system's performance when they are and are not enabled.

The DSA parameters being toggled in this phase of experimentation (i.e., "DSA All") are ***selfPreservation***, ***coordinateRepairs***, and ***effectorSensing***. All other parameters are fixed and can be found in Appendix B.

Ten replicates of each configuration were simulated with an identical set of 10 random seeds used to initialize all stochastic processes, including agent starting locations, walking, and asteroid impact times and locations. The generated data is averaged across all 10 runs during postprocessing. Note that the Figure of Merit (outcome variables) are all measured per time step during the simulations, so totals like number of asteroids or number of messages accumulate over the course of each run. As a result, both the rate of change and the final values attained provide meaningful performance metrics.

The results of the first phase of experimentation are shown in Fig. 3. Looking at the Avg. Pct. Asteroids Repaired (Fully Repaired Asteroids divided by Total Asteroids) graph, the performance improvement with the DSA-enabling parameters set to ON is quite salient. At 3000 ticks of the simulation, the average performance over 10 replicates of the DSA-enabled system shows that 90.4% of the damage inflicted by the asteroid impacts has been repaired, while the average performance over 10 replicates of the same set of random seeds of the system operating without those parameters enabled has repaired approximately 65.4% of the damage.

The Avg. Pct. Useful Work graph shows that with All DSA ON, the percentage of Useful Work, defined as:

$$\frac{Cumulative\ Necessary\ Repairs}{Cumulative\ Necessary\ Repairs + Cumulative\ Redundant\ Repairs}$$

produces an average of 85.6% at $t = 3000$. With All DSA OFF, the amount of useful repair work was approximately 35.8%.

Avg. Pct. Remaining Agents (Surviving Agents divided by Total Agents) shows the effect of toggling All DSA on average agent attrition. 79.7% of the agents survive with All DSA ON, and 67.4% survive with all DSA OFF.

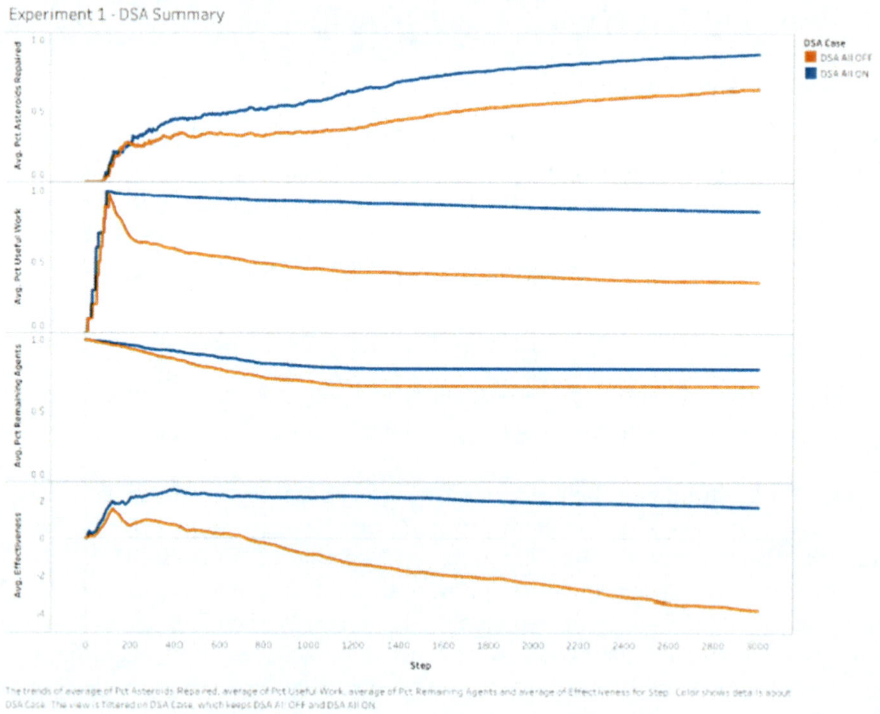

Fig. 3. Experiment 1 distributed situational awareness summary

Effectiveness in this context is defined as:

$$\frac{Cumulative\ Necessary\ Repairs - Cumulative\ Redundant\ Repairs}{Total\ Message\ Count}$$

The Avg. Effectiveness at $t = 3000$ with All DSA ON is 1.7. At $t = 3000$, with All DSA OFF, it is computed to be -3.8.

5.3 Experiment 2: Exquisite Few vs. Crude Many

Given the increase in performance measures seen with the introduction of the DSA parameters in Experiment 1, we fixed those to ON and turned our attention to the system's agent design trade space. For this phase of experimentation, we were interested in comparing performance across systems composed of few exquisite assets vs systems of many crude assets. The parameter combinations used for this experiment are shown in Table 1. Note that Design Point (DP) 5 is the baseline configuration used in Experiment 1.

Table 1. Experiment 2 exquisite few vs. crude many parameter combinations

Design point	Population size	Sensor range	Comms range	Effector range
1	30	162	108	54
2	60	144	96	48
3	90	126	84	42
4	120	108	72	36
5	150	90	60	30
6	210	72	48	24
7	240	54	36	18
8	270	36	24	12
9	300	18	12	6

Just as with Experiment 1, ten replicates of each configuration were simulated with an identical set of 10 random seeds used to initialize and realize all stochastic process outputs, including agent starting locations, walking decisions, and asteroid impact times and locations. The generated data is averaged across all 10 runs during postprocessing. DP 1 has the fewest agents with the largest range of capabilities per agent. DP 9 has the most agents and the lowest range of capabilities per agent. We were interested in which DP produced the most performant system in order to guide our design choices moving forward. It should be noted here that there is ongoing and future work which includes introducing different threat models into the simulation to ensure we are not inadvertently creating a system which is highly optimized to a single type of perturbation.

Figure 4 shows the results of Experiment 2. Looking at Avg. Pct. Asteroids Repaired, we see that DP 6 achieved the highest score at 91.2%, and DP 1 the lowest at 49.6%.

The Avg. Pct. Useful Work metric shows a tight cluster of final values across DPs. The lowest being DP 5 at 90.2%, and the highest being DP1 at 98.5%, with DP9 close behind at 98.5%. Avg. Pct. Remaining Agents graph shows another tight cluster of results. Regardless, DP1 came out on top with 72.3%, followed by DP8 at 70.6%. DP 5 came in last at 66.2% agents surviving.

The Effectiveness graph of Fig. 4 shows us a diverse set of results across system designs. DP9 stands out far above the rest with a result of 139.7. DP1 is second at 49.4 and DP8 third with 48.4. Interestingly, the rest of the DPs are clustered between DP4 at 2.7 and DP7 at 16.0.

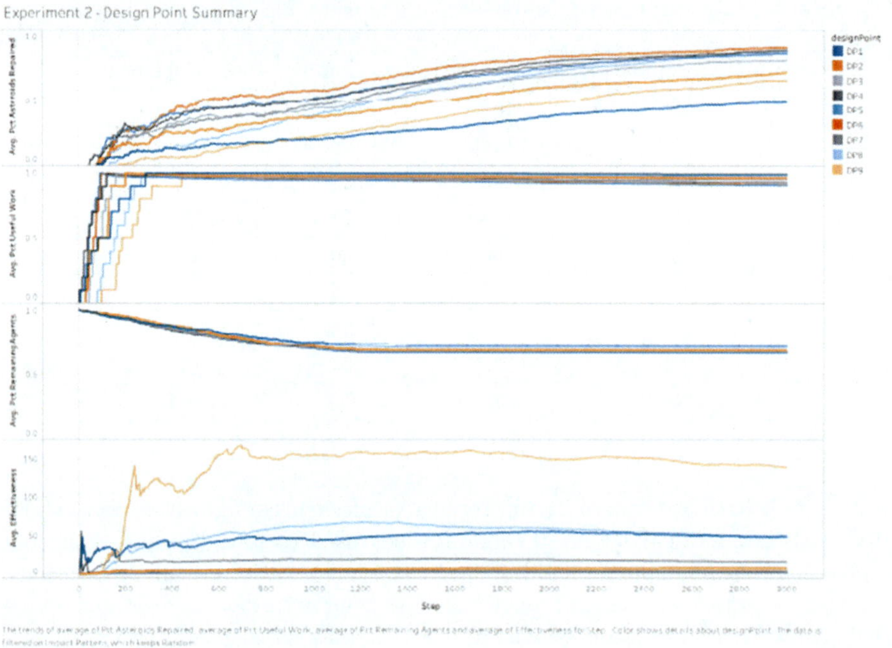

Fig. 4. Experiment 2 exquisite few vs. crude many summary

6 Discussion

We posit that functional emergence depends on DSA as well as decentralized decision-making. If situational awareness is not made available to an agent in the system, then the decisions being made by one agent will not be influenced by the previous decisions made by another agent. Although this seems to be obvious, there is sometimes a tendency to understate the importance of the embodiment and coordination-over-distance that more sophisticated, engineered complex adaptive systems may require.

It's important to keep in mind when evaluating a system that is functionally emergent that failures at the individual agent level do not necessarily indicate bad performance, as the dynamic whole is greater than the sum of the static parts. For example, expecting zero attrition is unreasonable in an uncertain and dangerous environment. Avoiding the potential for targeted attrition to lead to systemic ruin requires the development of systems that do not contain single points of failure and is one of the major motivators of taking on the challenge of mastering functional emergence despite the relative ease of creating centralized and optimized point solutions.

6.1 Experiment 1

The difference in the percentage of asteroid damage repaired when all DSA parameters were enabled is 25.0%. This indicates that imbuing our system with the foundational capabilities of DSA and being guided by the four pillars of functional emergence drove a strong improvement in the performance of the system.

The difference in the ratio of useful work to total work with the stigmergic parameter ***coordinateRepairs*** enabled is significant. Only 14.4% of all the repair work executed by the effector agents was unnecessary in the All DSA ON case. With a difference of 49.8% between All DSA ON and OFF, enabling indirect communication between leaf agents via embodied, real-time environmental sensing and decision-making has an incredible impact on system efficiency and performance.

In terms of survivability, we may conclude that 12.3% of agent attrition is caused by a lack of distributed situational awareness an agent has about its local environment and lack of decentralized decision-making capability that provides the optionality to temporarily override a self-destructive command.

The DSA-enabling parameters have a notable effect on the resulting Effectiveness metric. With a final difference of 5.4, and the All DSA OFF case value at −3.8, clearly the amount of redundant repairs was more than the number of necessary repairs, hence the value going negative, and then the magnitude of the negative value with All DSA OFF shows that the system's message passing was quite efficient compared to the amount of work performed (with the caveat that the large majority of the work was not useful to achieving the system-scale goal).

On the whole, performance across all three categories of measurement was improved with the activation of the three DSA parameters. To give us a way to summarize overall performance concisely, we averaged the values of Pct Asteroids Repaired, Pct Useful Work, and Pct Remaining Agents into a single overall metric. On this combined metric, ALL DSA ON ended up with a final score of 85.3% whereas ALL DSA OFF scored only 56.8%, yielding a 28.5% performance improvement using DSA.

6.2 Experiment 2

Using this same overall performance measure averaging the three individual percentage metrics, we found that DP 6 in Experiment 2 was the best performing design early and during most of the simulation run periods, with a final score of 82.9%. However, DP 8 ended up in the final best position, scoring slightly higher at 84.9%, for reasons discussed below. The most salient takeaway from this overall metric is that DP 1 is a terrible design, with the lowest of the final scores at 73.5%, and only barely able to repair half the asteroid-induced damage. Large, exquisite assets which are relatively few in number produce, in this case, a system which is the least capable of handling uncertainty. More work is needed to formulate conclusive statements, but the lesson taking shape here is that the enterprises designed with large, monolithic functional components are not the most performant in a stochastic and uncertain adversarial environment. Additionally, these DP 1-type systems leave little possibility for functional degeneracy.

The Avg. Pct. Useful Work metric reflects a tight cluster across DPs, which is most likely due to the effects of ***coordinateRepairs***, which was ON for the entire experiment, along with the other DSA parameters (see Appendix B for complete list of parameter settings). Interestingly, DP 1 had the highest score, barely ahead of DP 9. The difference between them amounts to 0.1%. Given the extreme ends of the spectrum these two DPs represent, the value of this metric in determining efficacy within this trade space is questionable. This measure is much more appropriate for determining the value of the DSA parameters as demonstrated in Experiment 1, but was nonetheless included here for consistency across the phases of experimentation.

The Avg. Pct. Remaining Agents results were not informative in determining robustness of the systems. In fact, it is quite surprising to see that relative survivability was not impacted by these highly differentiated design points. Further research is needed here, but some initial observations are that systems with larger populations present more targets for potential attrition and the simplifying assumptions that a more exquisite asset present the same targetable area and same susceptibility to damage as a crude one may need to be challenged.

Naïvely, the Effectiveness measure shows that DP8 and DP 9 were the most effective designs in terms of operational efficiency, with a huge amount of the work getting done with relatively little overhead communication costs. This seems counterintuitive considering that this DP has the largest number of individual agents, thus it would be safe to assume that the largest population would generate the most messages. Upon visual inspection of the simulation at runtime, it became clear to the research team that the reason for the apparent efficiency was that the comms range of the crudest population was so small, that very few links were getting established. It turned out that the functional degeneracy which enables the Effector agents to sense asteroid damage (parameter ***effectorSensing***) appeared to be wholly responsible for the system's functionality, despite the extremely limited range (an Effector's sensing range is equal to their effector range when ***effectorSensing*** is enabled). This means that the population of Effectors was large enough that they managed to stumble directly upon asteroid impacts during their random walking, which explains the relatively poor performance compared to most of the other DPs (65.6% asteroids repaired). It is intriguing that such a segmented and crude system functioned at all.

7 Next Steps

This research effort is just scratching the surface of decentralized system design. It is believed that the work presented here will feed further work defining decentralized system design principles. The results of this work have inspired a number of interesting next steps for this research.

7.1 Threat Modeling

The implicit belief that motivates the design of the system against a random scenario is that a system that can respond effectively under complete uncertainty will also be resilient in the face of less entropic threats. Currently, we are proceeding with implementing a number of threat models which go beyond randomly placed asteroid impacts to include a number of lower entropy patterns. Adding pseudo-intelligence to the threat model will allow us to evaluate if our system is overly-optimized for our random scenario. In essence, we want to try to avoid the creation of a "highly optimized tolerant", or HOT, system as described by [6]. Rather we would prefer to create a system more along the lines of a "COLD," constrained optimization with limited deviations [18], system. COLD systems have risk aversion in their design that dramatically reduce the likelihood of catastrophic failure. In order to move in that direction, we will define a number of different threat models to ensure we do not design for one type of threat at the expense of others.

Additionally, the size of impacts is currently homogenous via a fixed parameter. Implementing a ranged parameter for varying impact size within a given run will provide a more realistic threat across newly-developed impact patterns.

7.2 Deeper Analysis

Applying additional data analysis will allow us to understand how small changes to parameters may have nonlinear effects on outcomes.

We have inspected our results by averaging across replicates to characterize a given configuration. The mean may be misleading if the generated data has a non-gaussian distribution. Future efforts will apply more rigorous data analytics to ensure appropriate conclusions are being reached.

In the spirit of Experiment 1, further work is needed to tease apart which DSA parameter combinations produce maximum and minimum performance for which metrics. Certain combinations are suspected to be the most efficient yet least performant.

Extending the results of Experiment 2, a deeper dive into the capabilities endowed by varying the system Design Points is needed.

The long term vision we are working towards would be the identification of design principles that enable the creation of an enterprise whose tactical decisions are decentralized and whose functions are emergent, but is also capable of adapting itself on a fundamental level, i.e., not just enable agent-scale decisions from moment to moment (which are the subject of this prototypical effort and of most complex adaptive systems research), but also adapting the set of action(s) that an agent is afforded, the heuristics and/or policies that the agent follows when determining whether or not to take said action(s), and how/when they interact with their peers, etc. This may involve the introduction of some semblance of a hierarchical structure within which the decentralized components would be embedded or perhaps a decentralized governance mechanism for proposing rule changes and reaching consensus.

Acknowledgements. The authors would like to thank Alan Stone, Eric Bellows, Elizabeth Jones, Samuel Gomez, and Sarah Mulutzie for their inspiration and tireless support of this effort. Approved for Public Release; Distribution Unlimited. Public Release Case Number 21-2699.

NOTICE: Portions of this technical data were produced for the U. S. Government under Contract No. FA8702-19-C-0001 and W56KGU-18-D-0004, and is subject to the Rights in Technical Data-Noncommercial Items Clause DFARS 252.227-7013 (FEB 2014).

A ODD+D

A.1 Overview

Purpose. The purpose of this simulation is to explore design principles for an asteroid monitoring and response system, as a motivational problem for an autonomous system designed to achieve a goal against a largely unknown threat.

Entities, State Variables, and Scales. The model is made up of a set of simulated autonomous aerial vehicles, they are broken down into three types: sensors (these entities can sense specific events in the environment), communications nodes (these entities relay messages they hear), and effectors (these entities can impact their environment); the environment is made up of the Continental United States, time is relatively abstract in the simulation but a time step is roughly one hour of wall clock time and the simulation takes place over roughly 4 months.

Process Overview and Scheduling. The simulation progresses via time steps in the following general way: 1) if it is time for another asteroid impact, the effected patches' state are updated, 2) agents with the ability to sense the environment for asteroid impacts, 3) agents then process all messages received during the previous time step, 4) agent next push messages out to each other, 5) agents then move, 6) all agents engaged in damage mitigation next continue that work, 7) as agents have now moved, we next calculate the communications network, 8) finally, patches damaged by an asteroid "wane" (the thought here is that blast and fire damage will eventually dissipate on its own). Within each group, agent activation is randomized.

A.2 Design Concepts

Theoretical and Empirical Background. The theoretical and empirical background consists of the basic idea of emergence, in that the individual behaviors of a group of agents may produce a macroscopic pattern that is difficult to infer from a study of the components in isolation, or beyond the capabilities of the individual component. See the main body text for a complete discussion of the background on distributed situational awareness and on command and control.

Individual Decision-Making. Agents make a number of decisions in this simulation: a) they randomly adjust their heading when not engaged in other activities, b) they decide whether or not to respond to a message stating the location of an asteroid impact, c) when reacting to an asteroid impact they decide on a new speed, d) they decide whether or not to relay information to other agents, e) they decide if it is too dangerous to respond to an asteroid impact.

Learning. The simulated agents do not learn.

Individual Sensing. Agents in this model sense a number of features within their environment: a) their location (as they will not leave the airspace of the Continental US), b) their proximity to others so they will not crash into each other and can communicate with each other (communication ranges are finite), and c) damage caused by asteroids.

Individual Prediction. The simulated agents do not predict.

Interaction. The agents interact with each other via messages.

Collectives. There are a number of collectives within the simulation: Agents are of one of three types (effectors, sensors, and communications nodes), agents interact via communication networks which are dynamic and recalculated each time step.

Heterogeneity. Heterogeneity is part of the simulation in a number of ways, agent randomly adjust their heading when not responding to an asteroid impact, agents react to their environment and so they each have a unique trajectory over the course of the simulation, and agents are of one of three types.

Stochasticity. Stochasticity is used in a number of ways: we set the random seed and then build algorithmically a script for the asteroid impacts, we then reinitialize the random number generator and proceed to lay down the agents with random locations and headings.

More specifically: To vary the random behaviors described above, we include a seed parameter for the simulation's random number generator, and guarantee that if it is fixed, the same initial configuration of agents and patterns of asteroid impact will recur on repeated runs. The total number of initial agents present in the simulation is set by the ***numAgents*** parameter. The percent of those agents that are sensors and effectors (in addition to being comms agents) are set by the ***percentSensors*** and ***percentEffectors*** parameters respectively. Each agent type has parameters that define their reach radius for the behaviors they implement. The ***sensorRange*** parameter determines how far away from an asteroid impact sensors can be while still observing it for reporting purposes. The ***commsRange*** parameter represents the maximum signal propagation distance for being able to communicate with neighboring agents, and the ***effectorRange*** parameter determines how close an effector must be to an impact site to enable it to do repair work there. When agents are randomly walking, they vary their movement by making nominal changes to their current heading and speed, choosing a random normal increment (or decrement) from their current values. They are however, limited to a max velocity, as specified by the ***maxNodeSpeed*** parameter, and they will turn 180 degrees from their current heading if they are about to leave the boundary of the CONUS map. An implied minimum velocity of zero allows for agents to occasionally stand still.

Observation. Gathered data includes: messages sent, total damage caused by asteroid impacts, damage mitigated by agent activity, redundant agent mitigation activities, number of types of agents remaining in the simulation (agent are occasionally destroyed by an asteroid).

A.3 Details - Implementation

Initialization. Initialization is done in three distinct steps: 1) the random number generator is fixed and the asteroid script is created, 2) the raster image of the continental US is loaded to create the environment for the agents, 3) finally the sensors, communication nodes, and effector agents are created and randomly placed about the continental US. Specifically: the US map with an initial configuration can be found in Fig. 4, with 150 total agents, 10% of which are sensors shown as blue triangles with blue ***sensorRange*** circles, and 10% of which are effectors shown as red triangles with red ***effectorRange*** circles. The remaining comms-only agents are shown as green triangles, and the ***commsRange*** results in light grey links between agents who are close enough in space to communicate directly. Figure 2 shows the map display after several hundred steps during a simulation run. Asteroid impacts appear as X's, with A damage filled red circles that are in various states of dissipation, leaving them eventually as just red X's, turning to a magenta color when their B damage has been fully repaired. Effector agents who are on their way to an asteroid site, or who are stopped there to do repair work are shown in orange. The display uses these changes in size and color to visualize dynamic activities.

Input Data. The simulation uses raster data to define the continental US all other simulation data is created at initialization algorithmically.

Submodels. There are four basic submodels to this simulation. First, there is a submodel used for the asteroids. It is made up of three basic components: a) an algorithmic generator for the time and location of asteroid impacts, this was done to allow us to vary the script at initialization but also make it repeatable; b) a component that reads the script at runtime and creates the impact damage at the determined time and place; c) and finally a component that "degrades" the damage caused by the asteroid impact over time irrespective of agent activities.

More specifically: The random threat model is controlled by a few additional parameters, including some that determine how agents are affected by asteroid impacts. The ***maxAsteroids*** parameter determines how many asteroids will strike over the course of a simulation run, and they will fall at a uniform random rate determined by half the total simulation run time and the this max target. In the experimental runs described here, we set the ***maxTicks*** parameter to 3000 and ***maxAsteroids*** to 50, yielding an approximate 3.33% probability of an asteroid striking on any of the first 1500 steps in each run and then zero percent after that. The ***impactScale*** parameter is used to determine the radius of damage that an asteroid causes, which is currently fixed for all asteroids in a simulation run. There are two types of damage caused by an asteroid, one we call A damage which wanes over time via a constantly decreasing radius per time step until it dissipates completely. This type of damage can interfere with communications links that traverse the impacted area, such that there is a chance they will not be received normally. Furthermore, if any agent finds itself within an asteroid's A damage radius, it will die and be removed from the simulation. As a result, agents that happen to be near impact points will die, and those who try to enter a damaged area to quickly will also die. Two parameters affect these susceptibilities to A damage, the ***nodeTolerance*** parameter and the ***linkTolerance*** parameter. Asteroid A damage values greater than these settings will kill the agent or block the transmission link, respectively.

The other type of asteroid damage is called B damage, and this persists within the initial ***impactScale*** radius until effector agents arrive on the scene and begin to do their repairs. Each repairing effector can accomplish a linear amount of repair work per time step, clearing cells within the B damage radius. By default, the effector agents remain at an impact site until their sole efforts in repairing damage would be sufficient to clear the entire impact area. More intelligent divide and conquer collaborative repair efforts are described below as part of a DSA experimental parameter.

The second submodel is used to move messages among the agents. Messages are instantiated as agents for bookkeeping convenience. Message are created and passed around to provide information about the location of asteroids and who is responding to them. The third submodel is for agent movement, agents move in one of two ways: if they are responding to an asteroid impact they set their heading toward that asteroid impact location and move forward, if they are not responding to an asteroid impact they move randomly about the airspace above the continental US by adding a bit of noise to their current heading and speed (subject to a maximum and minimum speed). The final submodel is used by effector agents to repair damage caused by asteroids, here effector agents decrease the amount of damage associated with an asteroid impact once they have arrived on scene and do book keeping to measure the among of redundant work done on that particular asteroid impact.

B DOE Parameters

In order to accommodate the number of desired simulation runs as well as the high volume of generated data, we utilized a high performance computing cluster to perform the simulations. The cluster used is a collection of 32 compute nodes, each with 28 CPUs, making it useful for compute-intensive tasks. The large number of CPUs allows for tasks to be broken up into sub-tasks and run in parallel, thus making large compute jobs much more time efficient. We use the cluster in conjunction with NetLogo in order to run simulations using a wide variety of parameter combinations in a quick, efficient manner.

As described in Sect. 5, two experiments were run which varied different sets of parameters. Table 2 contains all of the parameters which were fixed in both Experiments 1 and 2 as well as their initial values.

Table 2. Parameters fixed across all experiments

Parameter name	Type	Value
Cluster delay	Int	25
Message protocol	Str	One time flooding
Variable dispatch speed	Bool	True
Max node speed	Float	1.5
Reverse	Bool	False
Percent effectors	Int	10
Percent sensors	Int	10
Fix seed	Bool	True
Cluster range	Int	50
Max ticks	Int	3000
Impact scale	Int	20
Link tolerance	Int	1000
Max asteroids	Int	50
Chance of asteroid	Int	5
Node tolerance	Int	5
Logging	Bool	None
Max dispatch distance	Int	100

Table 3 details the parameters varied in Experiment 1 and the set of possible values they could take on. Experiment 1 consisted of all combinations of the listed parameter values.

Table 3. Parameters for experiment 1

Parameter name	Type	Value(s)
Population size	Int	150
Sensor range	Int	90
Comms range	Int	60
Effector range	Int	30
Effector sensing	Bool	True, False
Self preservation	Bool	True, False
Coordinate repairs	Bool	True, False
Seed	Int	123, 456, 789, 987, 654, 321, 111, 222, 333, 444

The parameters contained in each of the 9 design points used in Experiment 2 are enumerated in Table 4. Experiment 2 consisted of all combinations of the 9 design points with the parameter values listed in Table 5.

Table 4. Design points used in experiment 2

Design point	Population size	Sensor range	Comms range	Effector range
1	30	162	108	54
2	60	144	96	48
3	90	126	84	42
4	120	108	72	36
5	150	90	60	30
6	210	72	48	24
7	240	54	36	18
8	270	36	24	12
9	300	18	12	6

Table 5. Parameters for experiment 2

Parameter name	Type	Value(s)
Effector sensing	Bool	True
Self preservation	Bool	True
Coordinate repairs	Bool	True
Seed	Int	123, 456, 789, 987, 654, 321, 111, 222, 333, 444

C Experimental Methodology

Our initial experiments with this ABM design demonstrated that the agents could, without any central C2, successfully detect asteroid impacts and repair them. However, the initial behaviors as described above did result in some pathological behaviors that we were able to discern by watching the simulation visualizations during manual runs. This section describes the iterative approach we used to propose ways to use various kinds of DSA, and the eventual experimentation we did to show the effects of using them or not.

The first pathology we observed was that the effectors were rushing to the scene of asteroids, often getting there before the killer type A damage had dissipated. Other agents simply doing their random walks were also being killed off in this manner, so we decided to give all the agents the ability to sense such danger in front of them, and to slow down or stall their movement if the path ahead would result in them being harmed. We parameterized this DSA information and associated behavior using a ***selfPreservation*** Boolean parameter.

The next pathology we observed was that many effectors were traveling great distances to reach an asteroid impact site and crowding around it, while later asteroids that fell near to where some of those effectors came from were going unattended for long periods of time. If the network connectivity was good (as it often was initially), the first asteroid could become known to most of the effector agents, and the unintended consequence of this was a swarming or bunching of effectors around these early asteroids. The first change was to enable effectors to preempt their targets when they were en route and knowledge of a closer target became available. While this helped some, it still made the enterprise too reactive, so we added additional behavioral logic for effectors that was controlled by a new parameter ***maxDispatchDistance***. The behavioral change from this was that when an effector was further than ***maxDispatchDistance*** from its closest known asteroid, it would continue to stay in its unassigned mission (or random walking) mode, and only dispatch to asteroids that were within ***maxDispatchDistance*** from its current position. All of the experiments we report on here had a fixed value for this parameter at a value of 100 distance units on the map.

The swarming or effector clumping behavior also enabled us to realize the benefit of letting a group of effectors collaborate on the repair work of a single asteroid (following the adage "many hands make light work"). In the real world, this would enable total repair time to be reduced significantly, so we modified the simulation to optionally allow agents to cooperate or not, via a DSA parameter called ***coordinateRepairs***, which was inspired by the Stigmergy principle since this would amount to agents being able to locally observe each other's work in the environment itself, and could be done without additional message passing overhead. When agents are coordinating, they can leave the scene of an asteroid impact as soon as all the B damage has been repaired by the collective effort, whereas they will operate as if they are alone if not cooperating, doing so-called "redundant" work and staying longer than they would otherwise need to. This redundant work effort is measured as a cost that offsets the necessary (or useful) work. Even when cooperating, however, agents who arrive at an asteroid that has been fully repaired are charged with a single time-step's worth of repair work as being redundant work, which might be saved if they had known not to keep coming when they were no longer needed. We anticipate that further kinds of DSA messaging regarding the state of asteroid repair can be used in later experiments to further reduce this kind of redundant work.

Another behavior we noticed was an obvious missed opportunity, when effector agents were on their way to an asteroid impact area and moved right past another one close to their current path that they hadn't been told about. Using the principle of functional degeneracy, we decided to optionally allow effector agents to also act as sensors, using their ***effectorRange*** reach radius parameter as a sensing range. This was controlled with a DSA parameter called ***effectorSensing***, and the behavior associated with it was to preempt the current mission target they were heading toward (if any) as well as to generate a Mission Information asteroid report and broadcast it to the network in the same manner that Sensor agents do.

In order to determine whether and how well our distributed agent design achieve enterprise mission objectives, we defined several performance metrics which alone or in combination could be used to evaluate any given parameterized scenario. All of these Figures Of Merit (FOMs) are measured dynamically at each simulation time step, so we can plot their incremental progress over time. All of the simulation experiments we describe below were performed with 3000 simulated time ticks, for two separate experiments that were designed to test particular hypotheses by varying certain of the parameters previously described. These FOMs represent our outcome variables in the experiments we ran, and they include agent attrition by type, percent of current total asteroids completely repaired, cumulative amount of repair work done (more on this later), and total messages transmitted by original sender agent type. We used these raw values to compute some aggregate measures as well, to combine the positive benefits with the negative costs. For example, we report on a metric called Effectiveness, which is computed as the necessary work minus the redundant work, all divided by the number of messages transmitted.

References

1. Alterman, R., Feinman, A., Introne, J., Landsman, S.: Coordinating representations in computer-mediated joint activities. In: Proceedings of the 23rd Annual Conference of the Cognitive Science Society, pp. 43–48. Lawrence Erlbaum Associates (2001)
2. Ashby, W.R.: Requisite variety and its implications for the control of complex systems. Cybernetica **1**(2), 83–99 (1958)
3. Axtell, R.: Three distinct kinds of empirically-relevant agent-based models. Brookings Institution Center on Social and Economic Dynamics Working Paper (2005)
4. Bonabeau, E., Hunt, C.W., Gaudiano, P.: Agent-based modeling for testing and designing novel decentralized command and control system paradigms. Technical report, Icosystem Corp, Cambridge, MA (2003)

5. Carley, K.M.C.: Coordinating for success: trading information redundancy for task simplicity, p. 10 (1990)
6. Carlson, J.M., Doyle, J.: Highly optimized tolerance: robustness and design in complex systems. Phys. Rev. Lett. **84**(11), 2529 (2000)
7. Ceruti, M.G.: Mobile agents in network-centric warfare. In: Proceedings 5th International Symposium on Autonomous Decentralized Systems, pp. 243–246. IEEE (2001)
8. Doursat, R., Ulieru, M.: Emergent engineering for the management of complex situations. In: Proceedings of the 2nd International Conference on Autonomic Computing and Communication Systems, pp. 1–10. Citeseer (2008)
9. Durfee, E.H., Lesser, V.R., Corkill, D.D.: Trends in cooperative distributed problem solving. IEEE Trans. Knowl. Data Eng. (1989)
10. Fuster, J.M., Bressler, S.L.: Past makes future: role of pFC in prediction. J. Cogn. Neurosci. **27**(4), 639–654 (2015). https://doi.org/10.1162/jocn_a_00746
11. Grimm, V., et al.: The odd protocol for describing agent-based and other simulation models: a second update to improve clarity, replication, and structural realism. J. Artif. Soc. Soc. Simul. **23**(2) (2020)
12. Hunt, E.R., Hauert, S.: A checklist for safe robot swarms. Nat. Mach. Intell. **2**(8), 420–422 (2020)
13. Kicinger, R.P.: Emergent engineering design: design creativity and optimality inspired by nature. Ph.D. thesis (2004)
14. Lee, T., Ghosh, S.: Simulating asynchronous, decentralized military command and control. IEEE Comput. Sci. Eng. **3**(4), 69–79 (1996)
15. Miller, J.H., Page, S.E.: Complex Adaptive Systems: An Introduction to Computational Models of Social Life. Princeton University Press, Princeton (2009)
16. Mulgund, S., Landsman, S.: User defined operational pictures for tailored situation awareness. In: Proceedings of the 12th International Command and Control Research and Technology Symposium-Adapting C2 to the 21st Century, Newport, RI, pp. 19–21. Citeseer (2007)
17. Müller, B., et al.: Describing human decisions in agent-based models-ODD+ D, an extension of the odd protocol. Environ. Modell. Softw. **48**, 37–48 (2013)
18. Newman, M.E., Girvan, M., Farmer, J.D.: Optimal design, robustness, and risk aversion. Phys. Rev. Lett. **89**(2), 028,301 (2002)
19. Norman, M.D.: Complex systems engineering in a federal IT environment: lessons learned from traditional enterprise-scale system design and change. In: 2015 Annual IEEE Systems Conference (SysCon) Proceedings, pp. 33–36 (2015). https://doi.org/10.1109/SYSCON.2015.7116725
20. Norman, M.D., Karavas, Y.G., Reed, H.: The emergence of trust and value in public blockchain networks. In: IX International Conference on Complex Systems, p. 22 (2018)
21. Norman, M.D., Koehler, M.T.K., Kutarnia, J.F., Silvey, P.E., Tolk, A., Tracy, B.A.: Applying complexity science with machine learning, agent-based models, and game engines: towards embodied complex systems engineering. In: Morales, A.J., Gershenson, C., Braha, D., Minai, A.A., Bar-Yam, Y. (eds.) ICCS 2018. SPC, pp. 173–183. Springer, Cham (2018). https://doi.org/10.1007/978-3-319-96661-8_18
22. Norman, M.D., Koehler, M.T., Pitsko, R.: Applied complexity science: enabling emergence through heuristics and simulations. In: Emergent Behavior in Complex Systems Engineering: A Modeling and Simulation Approach, pp. 201–226. Wiley (2018)
23. Berner, C., et al.: Dota 2 with large scale deep reinforcement learning. arXiv:1912.06680 [cs, stat] (2019)

24. Page, S.: The Diversity Bonus. Princeton University Press, Princeton (2019)
25. Railsback, S.F., Grimm, V.: Agent-Based and Individual-Based Modeling: A Practical Introduction. Princeton University Press, Princeton (2019)
26. Reynolds, C.W.: Flocks, herds and schools: a distributed behavioral model. In: Proceedings of the 14th Annual Conference on Computer Graphics and Interactive Techniques, pp. 25–34 (1987)
27. Ronald, E.M., Sipper, M.: Engineering, emergent engineering, and artificial life: unsurprise, unsurprising surprise, and surprising surprise. Artif. Life **7**, 523–528 (2000)
28. Salmon, P.M., Stanton, N.A., Walker, G.H., Jenkins, D.P., Rafferty, L.: Is it really better to share? Distributed situation awareness and its implications for collaborative system design. Theor. Issues Ergon. Sci. **11**(1–2), 58–83 (2010)
29. Scacchi, W., Brown, C., Nies, K.: Exploring the potential of virtual worlds for decentralized command and control. In: 17th International Command and Control Research and Technology Symposium (ICCRTS) (2012)
30. Taleb, N.N.: Antifragile: Things That Gain from Disorder, vol. 3. Random House Incorporated (2012)
31. Theraulaz, G., Bonabeau, E.: A brief history of stigmergy. Artif. Life **5**(2), 97–116 (1999)
32. Tisue, S., Wilensky, U.: NetLogo: a simple environment for modeling complexity. In: International conference on complex systems, vol. 21, pp. 16–21. Boston, MA (2004)
33. Ulieru, M., Doursat, R.: Emergent engineering: a radical paradigm shift. Int. J. Auton. Adapt. Commun. Syst. **4**(1), 39–60 (2011)
34. Van Dijk, H.: Situation awareness in crisis situations: development of a user defined operational picture. In: ISCRAM (2015)
35. Wilensky, U.: NetLogo: center for connected learning comp.-based modeling, vol. 158. Northwestern Univ, Evanston (1999)
36. Wilson, E.O.: Sociobiology: The New Synthesis. Belknap Press (1975)
37. Yang, A., Abbass, H.A., Sarker, R.: Evolving agents for network centric warfare. In: Proceedings of the 7th Annual Workshop on Genetic and Evolutionary Computation, pp. 193–195 (2005)

An Agent-Based Approach to Classical Competitive Prices

Jonathan F. Cogliano[1(✉)], Roberto Veneziani[2], and Naoki Yoshihara[3,4,5]

[1] Economics Department, University of Massachusetts Boston, Wheatley Hall, 5th floor, 100 William T. Morrissey Blvd., Boston, MA 02125, USA
jonathan.cogliano@umb.edu

[2] School of Economics and Finance, Queen Mary University of London, Mile End Road, London E1 4NS, UK
r.veneziani@qmul.ac.uk

[3] Department of Economics, University of Massachusetts Amherst, Crotty Hall, 412 North Pleasant Street, Amherst, MA 01002, USA
nyoshihara@econs.umass.edu

[4] The Institute of Economic Research, Hitotsubashi University, Naka 2-1, Kunitachi, Tokyo 186-8603, Japan

[5] School of Management, Kochi University of Technology, 2-22, Eikokuji-Chou, Kochi, Kochi 780-8515, Japan

Abstract. This paper develops a novel agent-based computational simulation of the classical competitive process and price formation. The simulation consists of many capitalists who make decisions of how to allocate capital across industries, and workers who sell their labour-power to capitalists. Capitalists make independent decisions of which good(s) to produce and engage in decentralised exchange with one another to obtain other goods needed for future production. Decisions of how to allocate capital across sectors are based on comparisons of capitalists' individual profit rates to industry averages. The ongoing migration of capital across industries captures key elements of the classical competitive process and formation of long-period competitive prices as centres of gravity for the movement of market prices. The simulation approach allows for a detailed study of the range of exchange dynamics and allocation decisions that are consistent with the formation of classical competitive prices, and adds further detail to the received theory of the classical long-period equilibrium.

Keywords: Agent-based model · Classical competition · Computational simulation · Prices of production

JEL Classification Codes: B51 · C62 · C63 · D51

1 Introduction

The classical political economists—notably Adam Smith and David Ricardo—and Karl Marx shared a theory of competitive prices and equilibrium, which

Z. Yang and E. von Briesen (Eds.): CSSSA 2021, SPCOM, pp. 152–163, 2022.
https://doi.org/10.1007/978-3-030-96188-6_11

views competitive prices as those consistent with uniform wage and profit rates across the economy. The formation of these competitive prices is typically explained as follows. Suppose one has a capitalist economy with a large number of industries populated by many workers and many capitalists. Workers sell their labour to capitalists for a wage and capitalists invest wealth and earn profit. Both workers and capitalists can move freely across industries in response to wage and profit rate differentials, respectively. The movement of workers and capitalists across industries will lead to the *tendential equalisation* of industry-level wage and profit rates: this is the *classical competitive process.* This competition and movement of workers and capitalists across industries is viewed as ongoing, causing persistent fluctuations in wage and profit rates, as well as market prices. The equilibrium wage, profit rate, and prices then emerge over a long period of time as *centres of gravity* for short-term market fluctuations.

The classical-Marxian notion of competition and process of price formation is often taken as sufficient justification for the stability of the corresponding long-period equilibrium, and use of competitive prices as an analytical tool. In fact, the typical formal approach in classical-Marxian economics relies solely on these long-period prices. However, the framework and process described above contain features of a complex system, notably, the open-ended dynamics of wage, profit rate, and price fluctuations, as well as the inherent degrees of freedom in the decentralised decisions of workers and capitalists. The similarities between complex systems and this classical-Marxian theory have been noted, in particular, by Foley (2003), Arthur (2006), Colander (2000), Matthews (2000), and Miller and Page (2007). The broad consistency between classical-Marxian theory and modern complexity approaches to economics, begs the question: are the processes of competition and open-ended oscillation of market prices around a long-period equilibrium something that can be studied using an approach more in-line with those used to analyse complex systems, particularly an agent-based computational model (ABM)?

This paper constructs an ABM along these lines to study the classical-Marxian process of price formation, and to scrutinise the stability of long-period prices. The model consists of many capitalists who make decisions of how to allocate capital across industries, and workers who sell their labour-power to capitalists. Capitalists make independent decisions on which good(s) to produce and engage in decentralised exchange with one another to obtain other goods needed for future production. Decisions on how to allocate capital across sectors are based on comparisons of capitalists' individual profit rates to industry averages. The ongoing migration of capital across industries captures key elements of the classical competitive process and formation of long-period competitive prices as centres of gravity for the movement of market prices.

Computational approaches to topics in classical-Marxian economies is a growing area of research, as covered in a recent survey by Cogliano et al. (2021a). Recent contributions by Jiang (2015), Cogliano and Jiang (2016), Wright (2008), and Cogliano et al. (2016; 2019; 2021b) all feature some type of computational simulation applied to topics in classical-Marxian economics. Cogliano et al. (2016; 2019; 2021b) analyse the equilibrium dynamics of exploitation and class in accumulation economies in a simulation environment in order to better understand: the conditions necessary for the persistence of exploitation and classes in the presence of accumulation (see Cogliano et al. (2016)); relationships between exploitation and skills, particularly the interaction of heterogeneous skills and inequalities in exploitation, wealth, and income (see Cogliano et al. (2019)); and the structure of exploitation and unequal exchange across countries (see Cogliano et al. (2021b)). These contributions yield insights in primary concerns of classical-Marxian economics, however, their focus is on equilibrium dynamics, rather than the process of equilibrium formation. Jiang (2015) and Cogliano and Jiang (2016) study Marx's circuit of capital and some aspects of classical-Marxian price and value theory. The latter, as well as recent work by Wright (2008), bear the greatest relevance to the analysis undertaken in this paper. These contributions construct ABMs of economies where labour is the only input of production to study the classical-Marxian process of gravitation around an equilibrium consistent with the labour theory of value (LTV), where prices are proportional to the labour time required for the production of goods.

Our approach here is distinct from earlier contributions in two key ways. First, the inclusion of both labour and capital in the production process, arguably, captures a more complete picture of the classical-Marxian framework. Second, our approach attempts to take minimal deviations from what can be considered rational behaviour. This provides a benchmark for analysing the stability of the long-period equilibrium, assuming conditions consistent with the rationality implied by the classical-Marxian competitive process. This ABM approach allows for a detailed study of the range of exchange dynamics and allocation decisions that are consistent (or inconsistent) with the formation of classical competitive prices, and adds further detail to the received theory of the classical-Marxian long-period equilibrium.

The rest of the paper proceeds as follows. Section 2 provides a brief overview of the standard formal treatment of classical-Marxian competitive prices. Section 3 presents the model. Section 4 presents results. Section 5 concludes.

2 Classical-Marxian Competitive Prices

The standard formal approach to classical-Marxian competitive prices assumes a linear circulating capital economy composed of n industries each producing a single good. There is a Leontief technology $(\mathbf{A}, \mathbf{L})$, where $\mathbf{A}$ is the $n \times n$ matrix of capital requirements with element a_{ij} denoting the amount of good i needed to produce one unit of good j, and $\mathbf{L}$ is the $1 \times n$ vector of direct labour requirements with element l_j. Prices are denoted by $\boldsymbol{p} = (p_1, \ldots, p_n)$, and w and r, respectively, denote the wage and profit rate. Competitive prices are then:

$$\boldsymbol{p} = (1 + r)(\boldsymbol{p}\mathbf{A} + w\mathbf{L}), \tag{1}$$

or

$$\boldsymbol{p} = (1 + r)\boldsymbol{p}\mathbf{A} + w\mathbf{L}. \tag{2}$$

The difference between Eqs. (1) and (2) is that the former assumes wages are paid prior to production, and the latter assumes wages are paid at the end of production.

This is the standard treatment of competitive prices in a classical-Marxian framework. One of the features of this approach is that, in addition to analysing prices for many goods, it is possible to derive a distributive relationship $w(r)$ to study the interactions of technical change and distribution between workers and capitalists, as well as the impact on prices.[1] This type of analysis is consistent with the aims of the Classicals and Marx. However, it leaves out the process through which equilibrium prices are formed, and the related gravitation of short-period prices around the equilibrium. The following section details an ABM designed to capture this gravitation process, while maintaining the above competitive prices as an important reference point.

3 The Model

Constructing an ABM of classical-Marxian competitive prices requires buiding a simulation consisting of many agents who, over successive periods of time, make decisions about producing goods, making exchanges, and allocating their productive capacity across multiple sectors of the economy. The details of this ABM are below. The economy is characterised by n goods, starting with the simplest case of $n = 2$, and time periods in the simulation are denoted by the subscript t. Similar to Sect. 2, the economy is linear and goods are produced using a Leontief technology $\mathbf{A}_t, \mathbf{L}_t$, where $\mathbf{A}_t$ is the same $n \times n$ matrix of capital requirements for production as in Sect. 2, and $\mathbf{L}_t$ is the $1 \times n$ vector of labour requirements for production, except $n = 2$. Consistent with the most common treatments of classical long-period prices, all capital circulates and there is no fixed capital.

Implementation of $\mathbf{A}_t, \mathbf{L}_t$ to produce goods, how the goods are exchanged, and how many producers operate in each sector is decided through the interaction of the many heterogeneous agents who populate the economy. For simplicity, we first assume that the technology is fixed over t and write $\mathbf{A}, \mathbf{L}$ for brevity.

3.1 Agents

Let $\mathcal{N}_t$ denote the set of agents at any time during the simulation with cardinality N_t. Individual agents in $\mathcal{N}_t$ are denoted by ν and the set of agents is partitioned into *capitalists*, $\mathcal{N}_t^c$, and *workers*, $\mathcal{N}_t^w$. $N_{i;t}^c$ denotes the number of capitalists in sector i at time t and N^w denotes the number of workers. Capitalists possess a $n \times 1$ vector of endowments of productive assets $\boldsymbol{\omega}_{t-1}^\nu = (\omega_{1;t-1}^\nu, \omega_{2;t-1}^\nu) \geq \mathbf{0}$ for all $\nu \in \mathcal{N}_t^c$. Capitalists aim to accumulate as much as possible, thus they reinvest all profits from $t-1$ at the beginning of t and do not consume.

[1] See Kurz and Salvadori (1995) and Cogliano et al. (2018) for exhaustive treatments of classical-Marxian competitive prices.

Workers supply the required amount of labour across the n sectors inelastically at the given subsistence wage. The subsistence wage is constant over t and denoted by the $n \times 1$ vector $\boldsymbol{b} > \mathbf{0}$, which represents a bundle that each worker needs to consume per unit of labour supplied. Workers do not save nor do they accumulate.

Wages are paid in advance of production and the $n \times n$ matrix of physical input requirements is given by $\mathbf{M} = \mathbf{A} + \boldsymbol{b}\mathbf{L}$. The $n \times 1$ vector $\boldsymbol{m}_i$ denotes the i-th column of $\mathbf{M}$ and describes the inputs a capitalist operating in sector i needs to produce one unit of good i. We assume that $\mathbf{M}$ is productive and indecomposable. Given that $n = 2$, the latter property implies that $m_{ij} > 0$ for any $i, j = 1, 2$. It is also assumed that $\mathbf{M}$ is specified such that the economy is able to produce a surplus during every t.[2]

3.2 Exchange

At the beginning of t, capitalists enter the market with $\boldsymbol{\omega}_{t-1}^{\nu}$ to exchange with other capitalists and ensure they have the goods necessary to undertake production. Exchanges between capitalists are bilateral and decentralised.

At the outset of t and the exchange procedure, capitalists are committed to operating in a particular sector $i = 1, 2$, regardless of the potential outcome of exchange. Let $\mathcal{N}_{i;t}^{c} \subset \mathcal{N}_t^c$ denote the subset of capitalists operating in sector i during t. $\nu \in \mathcal{N}_{i;t}^{c}$ enter the market with $\boldsymbol{\omega}_{t-1}^{\nu}$ and price expectations $\boldsymbol{p}_t^{e\nu} = (p_{1;t}^{e\nu}, p_{2;t}^{e\nu})$. $\boldsymbol{p}_t^{e\nu}$ are normalised such that $p_{1,t}^{e\nu} + p_{2;t}^{e\nu} = 1$, and are randomly determined at $t = 1$ with the constraint that $\nu \in \mathcal{N}_{i;t}^{c}$ expect to obtain a profit rate $r_{i;t}^{e\nu}$ that is nonnegative and higher than the profit rate in sector $j \neq i$. Formally:

$$r_{i;t}^{e\nu} = \frac{p_{i;t}^{e\nu} - \boldsymbol{p}_t^{e\nu}\boldsymbol{m}_i}{\boldsymbol{p}_t^{e\nu}\boldsymbol{m}_i}, \tag{3}$$

and, for all $\nu \in \mathcal{N}_{i;t}^{c}$, $r_{i,;t}^{e\nu} \geq 0$ and $r_{i;t}^{e\nu} \geq r_{j;t}^{e\nu}$, for $i \neq j$.

Given this, all $\nu \in \mathcal{N}_t^c$ operating in any sector i will want to produce as much of good i as possible subject to their wealth constraint. Thus, we can derive the vector of input demand functions $\boldsymbol{y}_t^{\nu} = y(\boldsymbol{p}_t^{e\nu}, \boldsymbol{\omega}_{t-1}^{\nu})$ for any ν. Because $r_{i;t}^{e\nu} \geq 0$, then $\boldsymbol{y}_t^{\nu} = \boldsymbol{m}_i z_{i;t}^{\nu}$ where $z_{i;t}^{\nu} = \frac{\boldsymbol{p}_t^{e\nu}\boldsymbol{\omega}_{t-1}^{\nu}}{\boldsymbol{p}_t^{e\nu}\boldsymbol{m}_i}$, or

$$\boldsymbol{y}_t^{\nu} = \boldsymbol{m}_i \frac{\boldsymbol{p}_t^{e\nu}\boldsymbol{\omega}_{t-1}^{\nu}}{\boldsymbol{p}_t^{e\nu}\boldsymbol{m}_i}. \tag{4}$$

From Eq. (4) it is possible to derive agents' trade functions $\boldsymbol{\tau}_t^{\nu} = \tau(\boldsymbol{p}_t^{e\nu}, \boldsymbol{\omega}_{t-1}^{\nu})$:

$$\boldsymbol{\tau}_t^{\nu} = \boldsymbol{m}_i \frac{\boldsymbol{p}_t^{e\nu}\boldsymbol{\omega}_{t-1}^{\nu}}{\boldsymbol{p}_t^{e\nu}\boldsymbol{m}_i} - \boldsymbol{\omega}_{t-1}^{\nu}. \tag{5}$$

For any agent ν with $\omega_{i;t-1}^{\nu} > 0$ and $\omega_{j;t-1}^{\nu} = 0$, for example, $\tau_{i;t}^{\nu} < 0$ and $\tau_{j;t}^{\nu} > 0$. Equation (5) is used to execute exchanges in Rule 1.

[2] For more on the typical treatment and requirements of input matrices in the classical-Marxian tradition, see Cogliano et al. (2018, Ch.2).

Rule 1 (Exchange). *The exchange algorithm proceeds as follows:*

1. *For each $\nu \in \mathcal{N}_t^c$, $\boldsymbol{\tau}_t^\nu$ is checked to determine which good agent ν demands and which they supply, i.e. which "side" of the market they are on, and $\mathcal{N}_t^c$ is partitioned into net demanders and net suppliers for good i. Let $N_{i;t}^c, N_{j;t}^c$ denote, respectively, the number of capitalists supplying goods i, j, respectively, during t according to $\boldsymbol{\tau}_t^\nu$.*
2. *In the case that $N_{i;t}^c \neq N_{j;t}^c$, each agent supplying good i is randomly matched with an agent μ supplying j. The subset $\mathcal{U}_t^c \subset \mathcal{N}_t^c$ denotes any agents who are are not matched at the start of exchange.*
3. *To be sure exchange is possible, for any pair of agents (ν, μ), it is checked that their endowments meet the condition:*
$$\min\left\{\frac{m_{11}}{m_{12}}, \frac{m_{21}}{m_{22}}\right\} < \frac{\omega_1^\nu}{\omega_2^\mu} < \max\left\{\frac{m_{11}}{m_{12}}, \frac{m_{21}}{m_{22}}\right\}. \tag{6}$$
Any pairs (ν, μ) satisfying (6) will then proceed with exchange. Any pairs (ν, μ) that do not satisfy (6) are added to $\mathcal{U}_t^c$.
4. *Prior to exchanges taking place, an attempt to rematch agents in $\mathcal{U}_t^c$ is made. $\boldsymbol{\tau}_t^\nu$ of agents in $\mathcal{U}_t^c$ are checked to partition the subset into sides of the market for good i: $\nu \in \mathcal{U}_t^c$ with $\tau_{i;t}^\nu < 0$ are matched with $\mu \in \mathcal{U}_t^c$ with $\tau_{i;t}^\mu > 0$. Condition (6) is checked again for pairs (ν, μ), and if satisfied, the rematched pairs move forward with exchange. If (ν, μ) do not satisfy (6), they remain in $\mathcal{U}_t^c$ and do not exchange during t.*
5. *The pairs (ν, μ) drawn from $\mathcal{N}_t^c \setminus \mathcal{U}_t^c$ then exchange at prices $\boldsymbol{p}_t^{(\nu\mu)} = (p_{1;t}^{(\nu\mu)}, p_{2;t}^{(\nu\mu)})$, such that $p_{1;t}^{(\nu\mu)} + p_{2;t}^{(\nu\mu)} = 1$, if and only if*
$$m_{ii}\frac{p_{i;t}^{(\nu\mu)}\omega_{i;t-1}^\nu}{\boldsymbol{p}_t^{(\nu\mu)}\boldsymbol{m}_i} - \omega_{i;t-1}^\nu = -\left(m_{ij}\frac{p_{j;t}^{(\nu\mu)}\omega_{j;t-1}^\mu}{\boldsymbol{p}_t^{(\nu\mu)}\boldsymbol{m}_j} - \omega_{i;t-1}^\mu\right). \tag{7}$$
If Eq. (7) holds for pairs (ν, μ), the pairs of agents exchange an amount of good i for an amount of good j determined by updating $\boldsymbol{\tau}_t^\nu$ for $\boldsymbol{p}_t^{(\nu\mu)}$, where $\boldsymbol{\tau}_t^{\nu\prime} = \boldsymbol{\tau}_t^\nu(\boldsymbol{p}_t^{(\nu\mu)}, \boldsymbol{\omega}_{t-1}^\nu)$ denotes the quantities of goods i and j traded.
6. *At the end of exchange, all agents $\nu \in \mathcal{N}_{i;t}^c$ update their endowments to reflect exchanges. Endowments after exchange are $\boldsymbol{\omega}_t^{\nu\prime} = \boldsymbol{\omega}_{t-1}^\nu + \boldsymbol{\tau}_t^{\nu\prime}$.*

Rule 1 relies on Walras's Law in executing trades. So long as there is no excess demand for good i, there will be no excess demand for good j either. Thus, $\boldsymbol{p}_t^{(\nu\mu)}$ are equilibrium prices for a given pair (ν, μ). Rule 1 is iterated until all pairs of agents who are able to exchange can complete one. After exchange is completed agents who completed a trade remember $\boldsymbol{p}_t^{(\nu\mu)}$ to update future price expectations, $\boldsymbol{p}_{t+1}^{e\nu} = \boldsymbol{p}_t^{(\nu\mu)}$, and to determine their profit rate, r_t^ν. All $\nu \in \mathcal{N}_t^c$ who were able to complete an exchange then proceed to produce during t.

Any agent η who was unable to complete an exchange during t is assigned placeholder prices $\boldsymbol{p}_t^{\eta\emptyset}$ such that $p_{j;t}^{\eta\emptyset} = \boldsymbol{p}_t^{\eta\emptyset}\boldsymbol{m}_j$, or $r_{j;t}^\eta = 0$. For any $\eta \in \mathcal{U}_t^c$ who did not exchange, placeholder prices are used to update price expectations during $t+1$, $\boldsymbol{p}_{t+1}^{e\eta} = \boldsymbol{p}_t^{\eta\emptyset}$. Agents who do not successfully complete an exchange during t do not produce during t.

3.3 Production

Production is carried out using $\mathbf{M}$ during every t by $\nu \in \mathcal{N}_t^c$ who successfully complete an exchange according to Rule 1. The production process during every t is formalised in Rule 2.

Rule 2 (Production). *During every t, each eligible $\nu \in \mathcal{N}_{i;t}^c$ operating in sector i uses $\boldsymbol{\omega}_t^{\nu\prime}$ and $\mathbf{M}$ to produce $x_{j;t}^{\nu} = 0$ units of $j \neq i$ and $x_{i;t}^{\nu}$ units of good i:*

$$x_{i;t}^{\nu} = \min\left\{\frac{\omega_{i;t}^{\nu\prime}}{m_{ii}}, \frac{\omega_{j;t}^{\nu\prime}}{m_{ji}}\right\}.$$

Let $\boldsymbol{x}_t^{\nu} = (x_{i;t}^{\nu}, x_{j;t}^{\nu})$ denote an agent's output vector and let $\boldsymbol{\omega}_t^{\nu\prime\prime}$ denote the endowment vector after production and prior to the end of t. At the end of production, $\boldsymbol{\omega}_t^{\nu\prime\prime} = \boldsymbol{\omega}_t^{\nu\prime} - x_{i;t}^{\nu}\boldsymbol{m}_i$, reflecting ν's endowments consumed in production. Endowments at the end of t are $\boldsymbol{\omega}_t^{\nu} = \boldsymbol{x}_t^{\nu} + \boldsymbol{\omega}_t^{\nu\prime\prime}$.

Any $\eta \in \mathcal{N}_t^c$ who did not exchange (nor produce) carry their endowments over to the next time period: $\boldsymbol{\omega}_t^{\eta} = \boldsymbol{\omega}_{t-1}^{\eta}$.

3.4 Capital Mobility

After production takes place $\nu \in \mathcal{N}_t^c$ evaluate their situation and decide whether to produce in a different sector during $t+1$ or remain in their current sector. This decision is based on agents' comparing their individual profit rates to the profit rate in the other sector. This comparison and the decision to switch sectors, or not, is made following Rule 3.

Rule 3 (Intersectoral Mobility). *At the end of t, every $\nu \in \mathcal{N}_{i;t}^c$ compares their realised profit rate, $r_{i;t}^{\nu}$, which is consistent with $\mathbf{p}_t^{(\nu\mu)}$, to the average profit rate in the other sector, $\bar{r}_{j;t}$, to evaluate whether to produce in sector j during $t+1$. For $j \neq i$, if $r_{i;t}^{\nu} \geq 0$, then every $\nu \in \mathcal{N}_{i;t}^c$ calculates:*

$$\psi_{j;t}^{\nu} = \frac{1}{1 + \exp\left[-\gamma\left(r_{i;t}^{\nu} - \bar{r}_{j;t}\right)\right]}, \tag{8}$$

where γ is a parameter influencing how strongly agents respond to differences in profit rates when making mobility decisions.

All $\nu \in \mathcal{N}_{i;t}^c$ then compare $\psi_{j;t}^{\nu}$ to a threshold $\rho_t^{\nu} \in [0, 1]$:

$$\begin{aligned}&\text{if } \psi_{j;t}^{\nu} \geq \rho_t^{\nu}, \nu \text{ switches to produce in sector } j \text{ during } t+1,\\ &\text{if } \psi_{j;t}^{\nu} < \rho_t^{\nu}, \nu \text{ continues producing in sector } i \text{ during } t+1.\end{aligned}$$

If for any $\nu \in \mathcal{N}_{i;t}^c$, $r_{i;t}^{\nu} < 0$, then ν automatically switches to produce in sector j during $t+1$. If $\nu \in \mathcal{N}_{i;t}^c$ switches sectors, then $\mathbf{p}_{t+1}^{e\nu} = \bar{\mathbf{p}}_{j;t}$, where $\bar{\mathbf{p}}_{j;t}$ corresponds to $\bar{r}_{j;t}$.

Switching sectors is costless for agents since all capital is circulating and no fixed capital investment, or divestment, is necessary. The decisions agents make about switching from one sector to another drive much of the dynamics of the simulation. Intersectoral mobility of this kind also captures the classical competitive process where capitalists continually seek higher returns on capital and migrate across sectors accordingly.

The sequence of exchange, production, and migration across sectors repeats for T time periods. The initialisation and parameters for the simulation are discussed below.

4 Simulations

Section 3 provides the general structure of the model and details of agents' behaviour. Below the initialisation conditions and parameters for the simulations are presented. Results from typical, sample simulations then follow.

4.1 Initialisation and Baseline Parameters

The initialisation of the simulation begins by first setting the number of goods $n = 2$ and determining the length of time T for which the simulation will run. At the start of the simulation the set of agents $\mathcal{N}_0$ is constructed with the number of capitalists, N_0^c, and workers, N_0^w determined and fixed for all $t = 1, \ldots, T$. Since workers are essentially passive in this version of the simulation, the crucial aspects of constructing $\mathcal{N}_0$ are in the initial conditions of the subset of capitalists $\mathcal{N}_0^c$ across the two sectors. At the start of the simulation, the number of capitalists operating in sector 1 is $N_{1;0}^c = 0.55N_0^c$, and conversely the number in sector 2 is $N_{2;0}^c = 0.45N_0^c$. Initial endowments for any $\nu \in \mathcal{N}_{i;0}^c$, $\boldsymbol{\omega}_0^\nu$, are such that $\omega_{i;0}^\nu > 0$ and $\omega_{j;0}^\nu = 0$. Thus, agents begin the simulation holding only the good for the sector to which they are initially assigned. The initial values for $\omega_{i;0}^\nu$ are randomly determined from a uniform distribution over the range $[0.2, 0.5]$. The ρ_t^ν agents use in evaluating switching sectors is constant across all t and uniform across all $\nu \in \mathcal{N}_t^c$, $\rho_t^\nu = \rho$, and γ from Eq. (8) is set equal to 1. These parameters and initial values, including the technology $(\mathbf{A}, \mathbf{L})$ and subsistence vector $\boldsymbol{b}$, are shown in Table 1.

The procedures in the simulation occur in the order of their presentation in previous subsections. The order, with time stages clearly marked, is summarised below:

1. $t = 0$: The simulation and agent set $\mathcal{N}$ are setup and initialised according to Table 1.
2. $t = 1$:
   (a) Based on initial conditions from $t = 0$, agents exchange according to Rule 1.
   (b) Eligible agents then produce according to Rule 2.
   (c) Agents evaluate switching sectors following Rule 3.

Table 1. Simulation parameters and initial values

Parameter/Variable	Baseline value
T	50
n	2
$\mathbf{A}_t = \mathbf{A}$	$\begin{pmatrix} 0.1\ 0.15 \\ 0.2\ 0.05 \end{pmatrix}$
$\mathbf{L}_t = \mathbf{L}$	$(0.165, 0.05)$
$\boldsymbol{b}_t = \boldsymbol{b}$	$(0.3, 0.5)$
N_0^c	100
$N_{1;0}^c$	$0.55N_0^c$
$N_{2;0}^c$	$0.45N_0^c$
$\omega_{i;0}^{\nu}$	$0.2 \leq \omega_{i;0}^{\nu} \leq 0.5$
$\omega_{j;0}^{\nu}$	0
$\rho_t^{\nu} = \rho$	0.9
γ	1

(d) Agents' endowments and price expectations are updated accordingly for $t + 1$.

3. $t > 1$: Repeat above items in step 2. until $t = T$.

The simulation is built and run in *Mathematica* version 12.

4.2 Results

This subsection provides an overview of our results. Figure 1 shows some sample results of the simulation. Figure 1(a) shows: relative exchange prices $p_{1;t}^{(\nu\mu)}/p_{2;t}^{(\nu\mu)}$ as dots in the diagram; relative average market price $\bar{p}_{1;t}/\bar{p}_{2;t}$ is the red line; and equilibrium relative price $\boldsymbol{p}^* = (1 + r)\boldsymbol{p}^*\mathbf{M}$ is the dashed horizontal line. Figure 1(b) shows the allocation of capitalists $N_{i;t}^c$, $N_{j;t}^c$ across the two sectors. Figures 1(c) and 1(d) show close-up images of prices over the course of the simulation, and Figs. 1(e) and 1(f) show the dynamics of average profit rates $\bar{r}_{1;t}$ and $\bar{r}_{2;t}$.

Figure 2 shows the same output for a different run of the simulation using the baseline parameters from Table 1. The results are qualitatively similar to those in Fig. 1, with some minor differences due to the randomness built into the simulation.

These results do not quite confirm the classical-Marxian theory of competition and price formation. In both Figs. 1 and 2, the average relative price oscillates as capitalists move between industries, but the equilibrium relative price consistent with a uniform profit rate does not act as a centre of gravity, or attractor, for average market price. These figures also show that many of the exchanges taking place between agents are happening at relative prices below

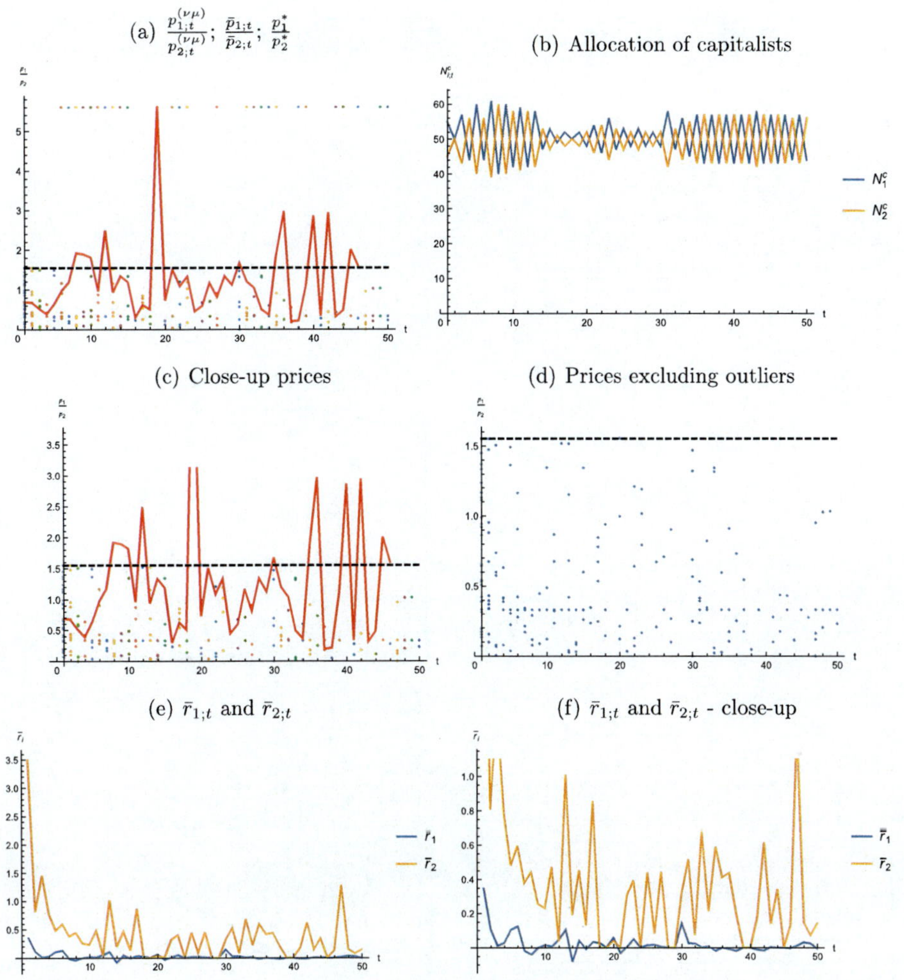

Fig. 1. Sample results

the equilibrium. One factor contributing to this is that, as Figs. 1 and 2 show, as profit rates exhibit fluctuations due to capitalists moving across industries, $\bar{r}_{2;t}$ remains higher than $\bar{r}_{1;t}$. Thus, the reallocation of capitalists across sectors is not generating the kind of response in prices that the classical-Marxian theory of competitive price formation would predict.

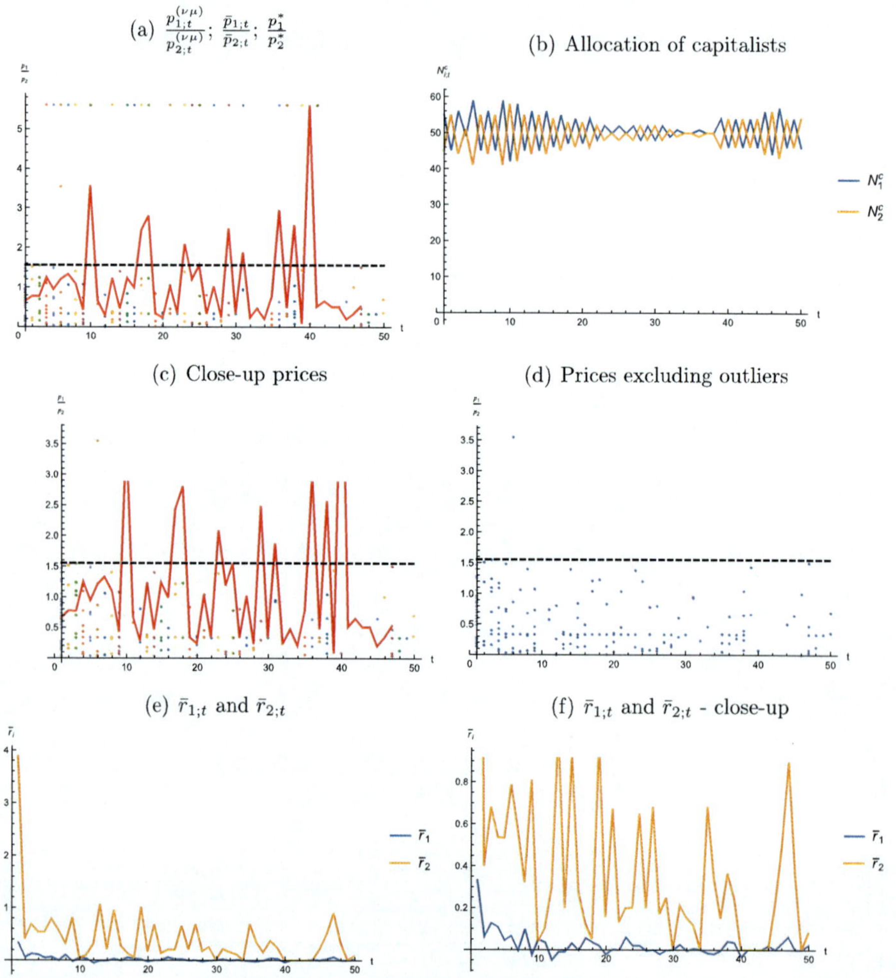

Fig. 2. Sample results II

5 Conclusions and Further Work

The results presented here are preliminary and a more detailed investigation into how the simulation responds to changes in parameters, initial conditions, and adjustments to the exchange and capital mobility procedures is still needed. It is possible that there exists a range of specifications of parameters and initial conditions for which the classical-Marxian process of price formation holds better.

There are also multiple avenues for extensions. For instance, it may be worthwhile to explore different exchange procedures, or to consider alternative ways for capitalists to evaluate switching sectors. It is also possible to incorporate either exogenous or endogenous technical change, to make workers non-passive, to allow capitalists to consume, and to consider adding more industries to the economy. These are just a sample of the range of possible refinements and extensions that exist. The current simulation provides a solid foundation for exploring these possibilities.

References

Arthur, W.B.: Out-of-equilibrium economics and agent-based modeling. In: Judd, K.L., Tesfatsion, L. (eds.) Handbook of Computational Economics, vol. 2. Elsevier B.V., Amsterdam (2006)

Cogliano, J.F., Flaschel, P., Franke, R., Fröhlich, N., Veneziani, R.: Value, Competition, and Exploitation: Marx's Legacy Revisited. Edward Elgar, Cheltenham (2018)

Cogliano, J.F., Jiang, X.: Agent-based computational economics: simulation tools for heterodox research. In: Lee, F.S., Cronin, B. (eds.) Handbook of Research Methods and Applications in Heterodox Economics. Edward Elgar, Cheltenham (2016)

Cogliano, J.F., Veneziani, R., Yoshihara, N.: The dynamics of exploitation and class in accumulation economies. Metroeconomica **67**(2), 242–290 (2016)

Cogliano, J.F., Veneziani, R., Yoshihara, N.: Exploitation, skills, and inequality. Rev. Soc. Econ. **77**(2), 208–249 (2019)

Cogliano, J.F., Veneziani, R., Yoshihara, N.: Computational methods and classical-Marxian economics. J. Econo. Surv. (2021a, forthcoming)

Cogliano, J.F., Veneziani, R., Yoshihara, N.: The Dynamics of International Exploitation. University of Massachusetts Boston, Economics Department Working Papers, WP N. 2021-02 (2021b)

Colander, D.: A thumbnail sketch of the history of thought from a complexity perspective. In: Colander, D. (ed.) Complexity and the History of Economic Thought. Routledge, London (2000)

Foley, D.K.: Unholy Trinity: Labor, Capital, and Land in the New Economy. Routledge, New York (2003)

Jiang, X.: Endogenous cycles and chaos in a capitalist economy: a circuit of capital model. Metroeconomica **66**(1), 123–157 (2015)

Kurz, H.D., Salvadori, N.: Theory of Production: A Long-Period Analysis. Cambridge University Press, Cambridge (1995)

Matthews, P.H.: Did Marx know the way to Santa Fe? In: Colander, D. (ed.) Complexity and the History of Economic Thought. Routledge, London (2000)

Miller, J.H., Page, S.E.: Complex Adaptive Systems. Princeton University Press, Princeton (2007)

Wright, I.: The emergence of the law of value in a dynamic simple commodity economy. Rev. Polit. Econ. **20**(3), 367–391 (2008)

Machine Learning Reveals Adaptive COVID-19 Narratives in Online Anti-Vaccination Network

Richard Sear[1(✉)], Rhys Leahy[1], Nicholas Johnson Restrepo[2], and Neil Johnson[1]

[1] The George Washington University, Washington, DC 20052, USA
{searri,neiljohnson}@gwu.edu
[2] ClustrX, LLC, Washington, USA

Abstract. The COVID-19 pandemic sparked an online "infodemic" of potentially dangerous misinformation. We use machine learning to quantify COVID-19 content from opponents of establishment health guidance, in particular vaccination. We quantify this content in two different ways: number of topics and evolution of keywords. We find that, even in the early stages of the pandemic, the anti-vaccination community had the infrastructure to more effectively garner support than their pro-vaccination counterparts by exhibiting a broader array of discussion topics. This provided an advantage in terms of attracting new users seeking COVID-19 guidance online. We also find that our machine learning framework can pick up on the adaptive nature of discussions within the anti-vaccination community, tracking distrust of authorities, opposition to lockdown orders, and an interest in early vaccine trials. Our approach is scalable and hence tackles the urgent problem facing social media platforms of having to analyze huge volumes of online health misinformation. With vaccine booster shots being approved and vaccination rates stagnating, such an automated approach is key in understanding how to combat the misinformation that slows the eradication of the pandemic.

Keywords: Topic modeling · Anti-vaccination network · Machine learning

1 Introduction

With declining vaccination rates despite widely available COVID-19 vaccines, it is crucial to understand how vaccine misinformation spreads online because misinformation can seed doubt and ultimately prevent people from trusting professional medical guidance. Moreover, the volume of misinformation on social media platforms poses challenges for content moderation teams and public health officials trying to raise credible information around vaccines. For example, Facebook's most recent quarterly report notes the ongoing issue of COVID-19 misinformation, despite their removing 20 M posts since the start of the pandemic [1].

The original version of this chapter was revised: The author "Yonatan Lupu" were included erroneously as co-author, which has now been corrected. The correction to this chapter is available at https://doi.org/10.1007/978-3-030-96188-6_14

Z. Yang and E. von Briesen (Eds.): CSSSA 2021, SPCOM, pp. 164–175, 2022.
https://doi.org/10.1007/978-3-030-96188-6_12

Furthermore, health misinformation the online battle for public opinion around vaccines began years before the COVID-19 pandemic [2]. A better understanding of such opposition ahead of vaccine rollouts is critical for scientists, public health practitioners, and governments. In particular, an automated (e.g. machine learning) approach could be ideal to help with moderation, given the huge amount of material posted each day.

Online social media platforms, and in particular the built-in communities that platforms like Facebook (FB) feature, have become popular fora for vaccine opponents (anti-vaccination) to congregate and share health (mis)information. Such misinformation can endanger public health and individual safety [3,4]. Likewise, vaccine supporters (pro-vaccination) also congregate in such online communities to discuss and advocate for professional public health guidance. Well before COVID-19, there was already an intense online conflict featuring anti-vaccination communities and pro-vaccination communities. Within anti-vaccination communities, the narratives typically draw on and generate misinformation about establishment medical guidance and distrust of the government, pharmaceutical industry, and new technologies such as 5G communications [3–5]. Adding fuel to this fire, the birth of the COVID-19 "infodemic" in January 2020 led to a plethora of misinformation on social media surrounding COVID-19 that directly threatens lives [6]. In one extreme case, a man died after consuming fish tank cleaner that contained chloroquine – a popular incorrect COVID-19 "cure" spread in online communities [7]. Unfortunately, the sheer volume of new online content and the speed with which it spreads, means that social media companies struggle to contain such health misinformation [8,9].

This study is motivated by: (1) the need for a deeper understanding of the intersection between online vaccination opposition and the online conversation surrounding COVID-19; and (2) the need for an automated approach since manual analysis is not viable at scale. We pursue an automated, machine learning approach that avoids the scalability limitations of manual content analysis. While the present paper is an early step in a challenging longer-term goal, the automated approach that we present allows the following questions to be addressed: how did COVID-19 guide conversations in anti-vaccination communities? What characteristics can machine learning reveal that imply the relative strength of the pro- and anti-vaccination communities' ability to attract new supporters?

Unlike many existing works, this study does not use Twitter data [10,11] since it is known that Twitter is more of a broadcast medium for individual shout-outs, whereas discussions and narratives tend to be nurtured in in-built online community spaces that are a specific feature of platforms like Facebook (e.g., fan page) [12]. Twitter does not have such in-built community spaces. In the present methodology, generalized from [13] and [14], data is collected from these online communities, specifically Facebook Pages that support either anti-vaccination or pro-vaccination views. This information is publicly available and does not require any individual details, just as understanding the content of a conversation among a crowd of people in an open, real-world public space does not require knowledge of any personal details about the individuals within that crowd. Details of our approach are given in the next section and the appendix. Though much work still needs to be done, this study is an early technical framework for a fully automated but interpretable understanding of the public health debate concerning vaccines and COVID-19.

2 Data and Machine Learning Methods

Facebook Pages, also known as fan pages or public pages, are accounts that represent organizations, causes, communities, or public figures. According to Facebook's policies, "Content posted to a Page is public and can be viewed by everyone who can see the Page" [15]. A Facebook Page is different from a Facebook personal account. Personal accounts represent private individuals, and their posts and interactions are considered more private and targeted to their immediate contacts. This paper does not analyze data from personal accounts. Our methodology follows [14] and [13] by analyzing the public content of Facebook Pages for both anti-vaccination ("anti-vax") and pro-vaccination ("pro-vax") communities. The publicly available content of these online communities is obtained using a snowball approach, starting with a seed of manually identified pages discussing either vaccines, public policies about vaccination, or the pro-vs-anti vaccination debate. Then their connections to other pages are indexed. At each step, new pages are evaluated through a combination of human coding and computer-assisted filters. To classify a page as being (1) anti-vaccination or pro-vaccination and (2) including COVID-19 content or not, we reviewed its posts and the page's "about" section. Pro-vaccination and anti-vaccination classifications required that either (a) at least 2 of the most recent 25 posts dealt with the pro-vaccination or anti-vaccination debate, or (b) the page's title or "about" section described it as pro-vaccination or anti-vaccination. At least two researchers classified each page independently. If they disagreed on their suggested classification, a third researcher reviewed the posts and then all three reviewers discussed these cases. Agreement was reached in each case. This also enabled us to distinguish between content that is intended to be serious versus merely satirical. The self-weeding tendency within Facebook Pages tends to reduce material from bots and fake profiles. We kept the present study focused on English, though other languages can easily be generalized using our procedure. Beyond that, our study was global and not limited to a particular region.

The content of these pages was then bundled together separately for the anti-vaccination community and the pro-vaccination community, and the two resulting sets of content were analyzed using machine learning. Specifically, we used an unsupervised machine learning technique called Latent Dirichlet Allocation (LDA) [16] to analyze the emergence and evolution of topics around COVID-19. The LDA method models documents as distributions of topics and topics as distributions of words. During the training process, these distributions are adjusted to fit the dataset. A more advanced version of LDA, dynamic LDA, also takes into account the time when the post was created, allowing it to model the evolution in word choice in particular topics as they change over time [17]. These methods are completely unsupervised; the only input required is a "number of topics" parameter, which determines the number of sets for the model to cluster text into. We use the Gensim implementation for these models.[1]

The coherence score provides a quantitative method for measuring the alignment of the words within an identified topic (see [18]) and is used here as a "goodness of fit" evaluation technique. It is generated from a separate algorithm which is run over a trained LDA model. The overall coherence score of a single model is the arithmetic mean of its per-topic coherences. There are many different coherence metrics to evaluate per-topic coherence. We use C_V which is based on a sliding window, one-set segmentation of the top words and an indirect confirmation measure that uses normalized pointwise mutual information and the cosine similarity. It comprises collections of probability measures on how often top words in topics co-occur with each other in examples of the topics. We refer to [18] for a full explanation and discussion of C_V. Our manual review of the word distribution making up each grouping showed that they do indeed correspond to reasonably distinct conversation topics. Details and examples are given in the Appendix.

Machine learning automation can, in principle, help address the issues facing social media platforms by mechanically picking out material that requires attention from the huge haystack of online content. Using both LDA methods (standard and dynamic) allows us to analyze different aspects of online narratives. Training many standard LDA models and evaluating them with C_V shows the evolution in *number of topics discussed* in short time frames, while training a longer-term dynamic LDA model shows the evolution of *important words within particular topics.*

3 Results and Discussion

We use several standard LDA models to examine the endogenous development of COVID-19 conversations at the beginning of the pandemic prior to the first officially reported U.S. COVID-19 death on February 29, 2020 [19]. We collected Facebook public post data for the period 1/17/2020–2/28/2020, inclusive. To assess the change over time, this period was divided into two time intervals of equal duration, T_1 and T_2 . The first time-interval, 1/17/2020–2/7/2020 (T_1), contains 774 total pro-vaccination posts and replies, and 3630 total anti-vaccination posts and replies. The second time-interval, 2/7/2020–2/28/2020 (T_2) contains 673 total pro-vaccination posts and replies, and 3200 total anti-vaccination posts and replies. We refer to [2] for a full discussion of this part of the study, but suffice it to say that even this early in the pandemic, the anti-vaccination narratives had an advantage for recruitment. Due to its tendency to discuss fewer topics as time progresses (see 1 and 2 [2]), the pro-vaccination community is less well-positioned to engage with the wide variety of blurry, and often more extreme, COVID-19 narratives that circulated during these time frames. With the anti-vaccination community's stronger ability to attract the attention

[1] https://radimrehurek.com/gensim/.

of the many different types of users who newly enter this online space in search of a particular nuanced 'flavor' of COVID-19 narrative that appeals to them, new users could be pulled toward the anti-vaccination cause. Combined with the distributed and decentralized structure of the anti-vaccination network (observed during these time frames as well as in later stages of the pandemic [20]), the anti-vaccination community had powerful recruiting capabilities even before vaccines were announced.

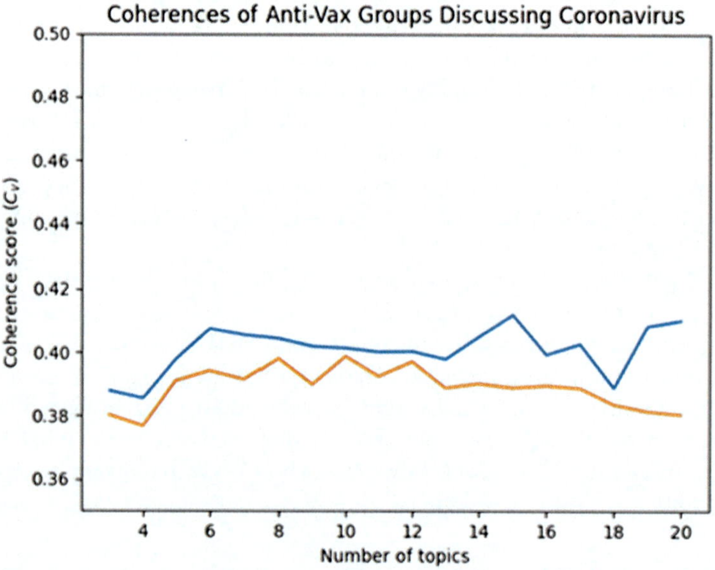

Fig. 1. Coherences of anti-vaccination groups (T_1 is blue, T_2 is orange)

In addition to this analysis, we also study anti-vaccination narratives over a much longer period of time using dynamic LDA. We trained models with their "number of topics" parameters ranging from 3–10. Figure 3 shows the average coherence scores for these models plotted over time. The best-performing model had 9 topics with an average C_V of 0.458. Figure 4 shows the individual coherence scores of topics within this model. Analysis of these individual topics provides a glimpse into how discussion within anti-vaccination groups evolved over the course of the first half of 2020.

Perhaps most interesting is Topic 5, which on human inspection of keywords is centered on reactions to lockdown orders. Over the course of the study period, this topic saw the largest change in C_V of any of the topics (increasing from 0.279 to 0.401), meaning that early on it did not contribute much to the model of online discussion, but rapidly became relevant. This makes intuitive sense as the first reported COVID-19 case in the U.S. was not until the end of February,

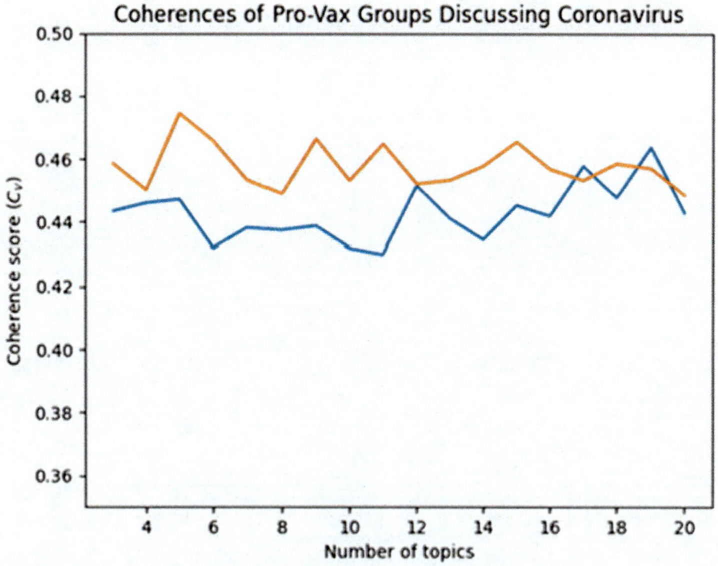

Fig. 2. Coherences of pro-vaccination groups (T_1 is blue, T_2 is orange)

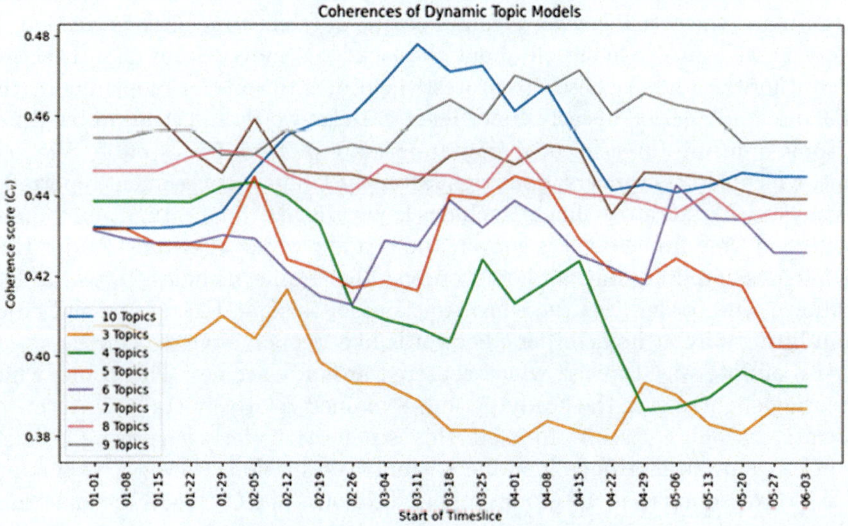

Fig. 3. Coherences of **all** anti-vaccination group dynamic topic models

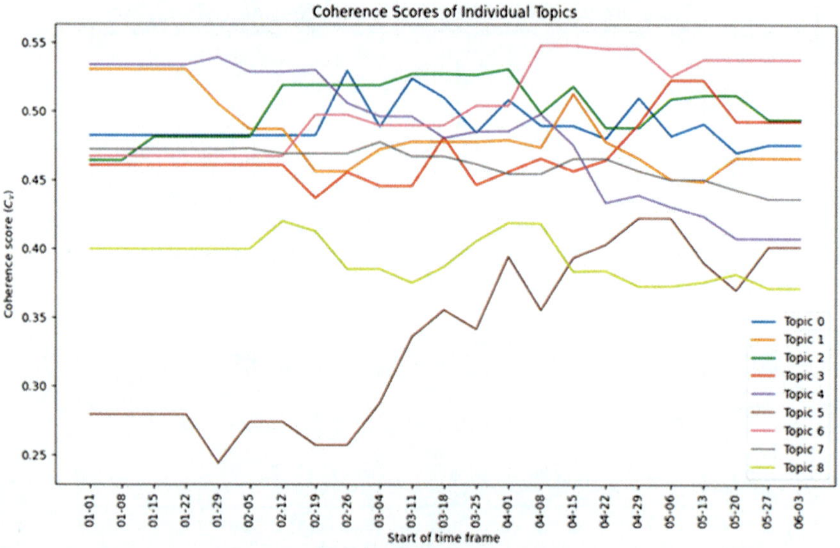

Fig. 4. Coherences of individual topics **within** the 9-topic dynamic topic model

and large-scale mitigation efforts did not appear until some weeks after that. Furthermore, the actual keywords found during different time periods within this topic suggest that conversation about the lockdown moved from dissatisfaction with authorities and the desire to protest the orders to calls for reopening by the end of our study period in early June. The top 10 keywords and their probabilities for Topic 5 during three selected time frames are plotted in 5. Notably, the first frame, which depicts conversations right after the Chinese government reported a new 'mystery' respiratory illness, includes keywords like 'quarantine' and 'home.' The second time frame depicts conversations a few weeks after most major U.S. major cities imposed stay at home orders; this frame includes keywords like 'lockdown' and 'order.' By the third time frame, as some U.S. states and cities began lifting stay at home orders, keywords like 'reopen' start to emerge.

Also of note was Topic 2, which centered around vaccines. This topic, while increasing slightly over the study period, remained relatively stable in terms of its relatively high C_V score. In itself, this is not particularly surprising because the posts came from groups identified as anti-vaccination. However, examining the keywords shows that the groups followed rollouts of COVID-19 vaccine trials, with "moderna" appearing in the top keywords in later time frames.

Another consistently coherent topic was Topic 6, which centered around video sharing. This topic began as the fifth most coherent topic (by a tight margin, see 4) and ended as the most coherent topic, gaining prevalence as COVID-19-related videos became more frequent within these Facebook groups. Because a keyword that emerges over time within this topic is "interview," the text surrounding these video links seems to indicate that the anti-vaccination network was interested in 'expertise' on the virus. A simple keyword search within posts

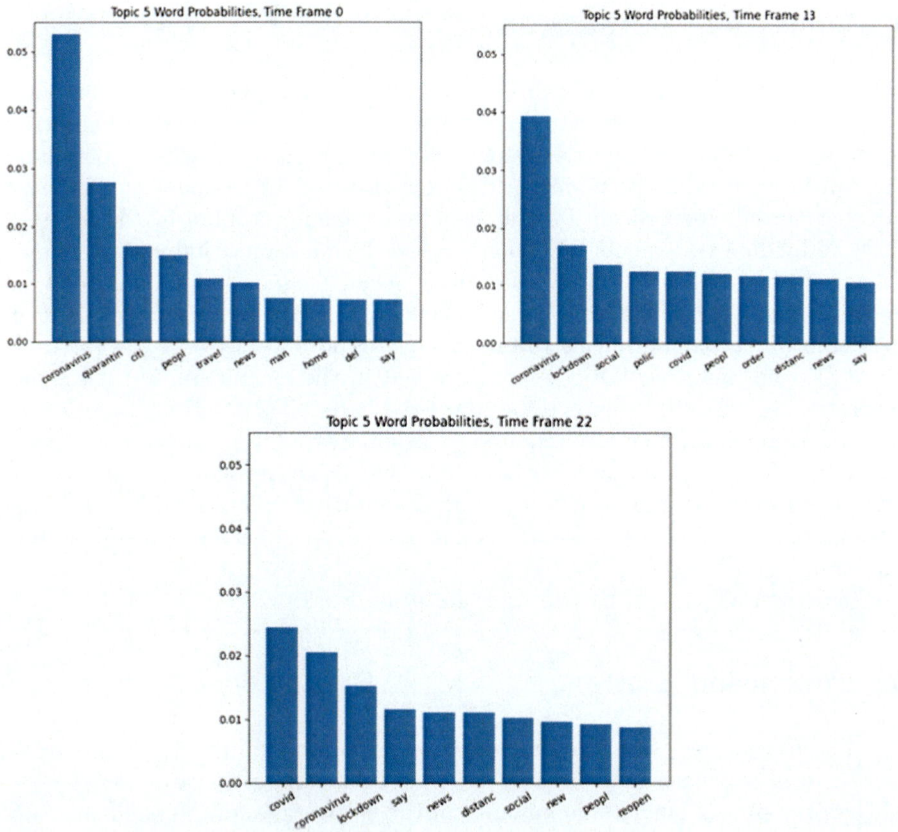

Fig. 5. Keywords in Topic 5 at selected time frames. Time Frames 0, 13, and 22 correspond to 1/1/20, 4/1/20, and 6/3/20 respectively. These time frames represent the first frame in the study period, a middle frame in a local C_V peak (see 4), and the last frame in the study period.

strongly connected with this topic reveals that frequently, members of the anti-vaccination network circulate third-party websites in addition to YouTube links, all featuring a small handful of "experts" who appear frequently. However, the methodology presented here cannot determine in an automated fashion who was interviewed or whether they were legitimate scientific authorities. Such a question is left for future research.

This finding also underscores the multiplatform nature of the anti-vaccination network and the importance of strong content moderation all over the internet, notably on video-sharing websites such as YouTube. From our methods, it is not clear whether the videos shared were themselves pro- or anti-vaccination, but it is clear that media content external to Facebook was an important aspect of the narratives.

4 Limitations of the Study

There are various limitations of this study. For example, other social media platforms apart from Facebook should be explored; however, we note that Facebook is the largest. Similar behaviors likely arise on any platform where communities can form. It will also be interesting, for example, to compare our findings to other studies focused on Twitter, where messaging is more in the form of short, individual statements [11]. There is also the question of influence of external agents or entities [10]. However, these social media communities tend to police themselves for bot-like or troll behavior. Further analysis is required of the details of the content. This will require going beyond just text and LDA; as we mentioned, multimedia posts are prevalent in these communities. Also, the generative model output needs to be compared in detail to the time-evolution of topics. Further research is also required to formulate findings across all platforms into actionable results for policy makers. It would also be a useful exercise to analyze pro-vaccination and potentially neutral narratives using the dynamic LDA model and compare the evolution of conversation topics between corresponding time frames.

These limitations will be addressed in future work.

5 Conclusion

Our findings suggest that the online anti-vaccination community developed a more diverse and hence more broadly accommodating discussion around COVID-19 than the pro-vaccination community in the early stages of the pandemic. As a result, the pro-vaccination community was less engaging to the heterogeneous ecology of potential new users who join the online COVID-19 discussion, and who may arrive online with a broad set of concerns, questions, and possibly preconceived notions, misinformation and even falsehoods.

The analysis in this paper also provides a first step toward eventually either replacing or supplementing the non-scalable efforts of human moderators tasked with identifying online misinformation.

Overall, this approach shows that a machine-learning algorithm (LDA) identifies plausible topics within collections of posts from online communities surrounding the vaccine and COVID-19 debate and uncovers details within the narratives that could be useful to content moderators, researchers, or policy makers. In addition to being able to handle large quantities of data, its results emerge quickly using statistical grouping techniques, instead of having to rely on potentially biased, slow, and costly human labeling.

Acknowledgement. CrowdTangle data are made available through The George Washington University. We are grateful for funding for this research from the U.S. Air Force Office of Scientific Research under award numbers FA9550-20-1-0382 and FA9550-20-1-0383.

6 Appendix

As mentioned in the main text, the methodology starts with a seed of manually identified Facebook Pages discussing either vaccines, public policies about vaccination, or the pro-vs-anti vaccination debate. Then their connections to other fan pages are indexed. At each step, new findings are vetted through a combination of human coding and computer assisted filters. This snowball process is continued, noting that new links can often lead back to members already in the list and hence some form of closure can in principle be achieved. This process leads to a set containing many hundreds of pages for both the anti-vaccination and pro-vaccination communities. Before training the LDA models, several steps are employed to clean the content of these pages in a similar way to other LDA analyses in the literature:

1. Mentions of URL shorteners are removed, such as "bit.ly". These are fragments output by Facebook's CrowdTangle API.
2. Many of the posts link to external websites. The fact that these specific websites were mentioned could itself be an interesting component of the COVID-19 conversation. Hence instead of removing them completely, the pieces ".gov", ".com", and ".org" were replaced with "__gov", "__com", and "__org", respectively. This operation effectively concatenates domains into a form that will not be filtered out by the later preprocessing steps.
3. The posts are then run through Gensim's `simple_preprocess` function, which tokenizes the post on spaces and removes tokens that are only 1 or 2 characters long. This step also removes numeric and punctuation characters.
4. Tokens that are in Gensim's list of stopwords, are removed. For example, "the" is not a good indication of a topic.
5. Tokens are lemmatized using the WordNetLemmatizer from the Natural Language Toolkit NLTK, which converts all words to singular form and/or present tense.
6. Tokens are stemmed using the SnowballStemmer from NLTK, which removes affixes on words.
7. Any remaining fragments of URLs (other than domain) that are left over after stemming, such as "http" and "www", are removed.

Steps 5 and 6 help ensure that words are compared fairly during the training process, and that if a particular word is a strong indicator of a topic, its signal is not lost just because it is used in many different forms. These steps rely on words existing in NLTK's pretrained vocabulary. Any word not in this vocabulary is left unchanged. After this preprocessing, we then train the LDA models on the cleaned data. We refer to [2] for a complete discussion of the standard LDA models employed. 8 dynamic LDA models were trained with their "number of topics" parameter ranging from 3–10 (inclusive) and each time frame consisting of the data gathered from the anti-vaccination groups in 1-week periods. While the amount of data available in each time frame is not uniform, we believe there is sufficient data in each time frame for the model to make useful inferences.

The code used to run our experiments is available and documented here: https://github.com/gwdonlab/topic-modeling. It is meant as a framework that can be used to run similar experiments on any text dataset.

References

1. Community standards enforcement report, second quarter 2021. About Facebook, 18 August 2021. https://about.fb.com/news/2021/08/community-standards-enforcement-report-q2-2021/. Accessed 15 Sep 2021
2. Sear, R.F., et al.: Quantifying COVID-19 content in the online health opinion war using machine learning. IEEE Access **8**, 91886–91893 (2020). https://doi.org/10.1109/ACCESS.2020.2993967
3. Larson, H.J.: Blocking information on COVID-19 can fuel the spread of misinformation. Nature **580**(7803), 306–306 (2020). https://doi.org/10.1038/d41586-020-00920-w
4. Kata, A.: A postmodern Pandora's box: anti-vaccination misinformation on the internet. Vaccine **28**(7), 1709–1716 (2010). https://doi.org/10.1016/j.vaccine.2009.12.022
5. Coronavirus: scientists brand 5G claims 'complete rubbish,' BBC News, 15 April 2020. https://www.bbc.com/news/52168096. Accessed 03 Sep 2021
6. Mythbusters. https://www.who.int/emergencies/diseases/novel-coronavirus-2019/advice-for-public/myth-busters. Accessed 03 Sep 2021
7. A man thought aquarium cleaner with the same name as the anti-viral drug chloroquine would prevent coronavirus. It killed him. Washington Post. http://www.washingtonpost.com/nation/2020/03/24/coronavirus-chloroquine-poisoning-death/. Accessed 16 Sep 2021
8. Frenkel, S., Alba, D., Zhong, R.: Surge of virus misinformation stumps Facebook and Twitter. The New York Times. 08 Mar 2020. https://www.nytimes.com/2020/03/08/technology/coronavirus-misinformation-social-media.html. Accessed 03 Sep 2021
9. Iyengar, R.: The coronavirus is stretching Facebook to its limits CNN. https://www.cnn.com/2020/03/18/tech/zuckerberg-facebook-coronavirus-response/index.html. Accessed 03 Sep 2021
10. Broniatowski, D.A., et al.: Weaponized health communication: twitter bots and Russian trolls amplify the vaccine debate. Am. J. Public Health **108**(10), 1378–1384 (2018). https://doi.org/10.2105/AJPH.2018.304567
11. Lama, Y., Chen, T., Dredze, M., Jamison, A., Quinn, S.C., Broniatowski, D.A.: Discordance between human papillomavirus Twitter images and disparities in human papillomavirus risk and disease in the United States: mixed-methods analysis. J. Med. Internet Res. **20**(9), e10244 (2018). https://doi.org/10.2196/10244
12. Ammari, T., Schoenebeck, S.: Thanks for your interest in our Facebook group, but it's only for dads': social roles of stay-at-home dads. In: Proceedings of the 19th ACM Conference on Computer-Supported Cooperative Work and Social Computing, New York, pp. 1363–1375, February 2016. https://doi.org/10.1145/2818048.2819927
13. Johnson, N.F., et al.: Hidden resilience and adaptive dynamics of the global online hate ecology. Nature **573**(7773), 261–265 (2019). https://doi.org/10.1038/s41586-019-1494-7

14. Johnson, N.F., et al.: New online ecology of adversarial aggregates: ISIS and beyond. Science, June 2016. https://www.science.org/doi/abs/10.1126/science.aaf0675. Accessed 03 Sep 2021
15. Facebook. https://www.facebook.com/policies_center/pages_groups_events. Accessed 03 Sep 2021
16. Blei, D.M., Ng, A.Y., Jordan, M.I.: Latent Dirichlet Allocation, 30 (2003)
17. Blei, D.M., Lafferty, J.D.: Dynamic topic models. In: Proceedings of the 23rd International Conference on Machine Learning - ICML 2006, Pittsburgh, Pennsylvania, pp. 113–120 (2006). https://doi.org/10.1145/1143844.1143859
18. Syed, S., Spruit, M.: Full-text or abstract? examining topic coherence scores using latent dirichlet allocation. In: 2017 IEEE International Conference on Data Science and Advanced Analytics (DSAA), pp. 165–174, October 2017. https://doi.org/10.1109/DSAA.2017.61
19. CDC newsroom, CDC, 01 January 2016. https://www.cdc.gov/media/releases/2020/s0229-COVID-19-first-death.html. Accessed 03 Sep 2021
20. Johnson, N.F., et al.: The online competition between pro-and anti-vaccination views. Nature **582**(7811), 230–233 (2020). https://doi.org/10.1038/s41586-020-2281-1

Agentization of Two Population-Driven Models of Mathematical Biology

John C. Stevenson(✉)

Doctor of Philosophy in Mechanical Enineering, Long Beach Institute, Long Beach, NY, USA
jcs@alumni.caltech.edu

Abstract. Single species population models and discrete stochastic gene frequency models are two standards of mathematical biology important for the evolution of populations. An agent based model is presented which reproduces these models and then explores where these models agree and disagree under relaxed specifications. For the population growth models, attainment of equilibrium based on constant population size is misleading for underlying population characteristics which may be nowhere close to equilibrium. These models also require homogeneous mixing which prevents prediction of extinctions due to local resource depletion. The discrete stochastic gene frequency models assume well mixed populations at constant size under neutral and weak selection pressure. The models' predictions for non-constant populations in strongly oscillating and chaotic regimes are surprisingly good, only diverging from the ABM at the most chaotic levels.

Keywords: Agentization · ABM · Single-species logistic growth equations · Discrete stochastic gene frequency models · Population-driven models · Weak selection pressure

1 Introduction

Mathematical biology provides many equation-based models of single species population dynamics that are important for evolutionary processes, two of which are: single species logistic equations (both continuous and discrete), and discrete stochastic models of gene frequencies in finite populations. The discrete logistic equations generate oscillating and chaotic population dynamics, and allow for some extinctions [2]. These discrete logistic growth models have a "surprising wide spectrum" of uses: tumor cell growth, fishery management [2], aggregation of phytoplankton [10], monocarpic plants, semelparouas insects and fish [3], and habitat loss and fragmentation [13]. The discrete stochastic gene frequency models are important for animal husbandry, trait extinction, genetic bottlenecks and, with some of these models, retrospective studies of allele frequency distributions based on current conditions [4,5].

Z. Yang and E. von Briesen (Eds.): CSSSA 2021, SPCOM, pp. 176–189, 2022.
https://doi.org/10.1007/978-3-030-96188-6_13

These models represent the "struggle for existence" which is an endogenous driver of evolution [7]. Any agent based model (ABM) that evolves agent or population characteristics should reproduce these standard results. An ABM is presented that reproduces these models' results and then compares the behaviors of these standard models to the ABM when the specifications are relaxed. The ABM modeling of the logistic growth equations reveals two important behaviors not seen in the standard models: spatially driven local extinctions and non-equilibrium population characteristics extending orders of magnitude past the time that the population size itself achieved equilibrium. The ABM modeling of gene frequencies reproduces the predictions of fixation probability and mean absorption times of the standard discrete stochastic models across a wide range of intrinsic growth parameters, up to the edge of chaotic population trajectories. These parameters are investigated for both neutral and weak selection pressures.

The relaxed specifications of the ABM generated differences with the standard models that highlight current research in spatial effects on both gene frequency models [9,10,13] and logistic growth models [8,11,12]; and in variable population sizes on logistic growth models [5]. That these areas of research address the differences identified by the ABM provide added confidence in the ABM modelling approach and this specific implementation. The equation-based research cited is driven by a desire for "mechanistic understanding" [8] and ability to "generalize" [9] that avoids "computationally intensive solutions" [8] and "highly complex computer simulation models" [10]. All of these equation-based approaches employ approximations that require "some fairly heavy analytical machinery" [9] and are tested against either ABMs or other numeric solutions of the stochastic, equation-based models developed.

The replicating ABM is based on the model of Epstein and Axtell [1,18]. The model uses identical agents on an equal opportunity (flat) landscape to match the standard models. For the logistic growth equations, as in the study of bacteria [7], the population trajectory begins with a single agent and is dependent on a stochastic growth parameter: infertility. There is a one-to-one mapping of the infertility parameter of the ABM with the intrinsic growth rate of the single species logistic growth equations. Detailed descriptions of the ABM parameters and process are given in Appendix A. The mathematics of continuous and discrete logistic growth equations used in this study are given in Appendix B.

This ABM, with the addition of a single gene with two alleles, allows replication of two Wright Fisher class discrete stochastic models of gene frequency. The probability of allele fixation and mean absorption times are replicated for the Cannings [21] and Moran [22] models of two allele haploid populations of fixed size.

2 Discrete Logistic Equation with Time Delay

The fields of mathematical biology and ecology developed equation-based modeling of single species populations, models both continuous and discrete [2,3]. A continuous homogeneous model of single species population was proposed by

Verhulst in 1838 [6] and allows an exact solution for the intrinsic growth rate. While the continuous model fits the initial phase of growth well, it does not model oscillating population sizes at high rates of intrinsic growth. These types of periodic and chaotic oscillations are often generated by population models that are discrete, that contain time delays, or both [24]. Researchers in the fields of biology and ecology have used these discrete and delayed population models to handle, for example, species that have no overlap between generations [2] or have specific breeding seasons [3]. To account for such delays in animal populations, Hutchinson in 1948 [20] and Wright in 1955 [19] extended the Verhulst process, now often referred to as the Hutchinson-Wright equation, as [3]

$$N(t+1) = [1 + r - \frac{N(t-\tau)}{K}]N(t) \tag{1}$$

with an explicit time delay τ in the self-limiting term. For the ABM, the time delay captures both the landscape cell's regeneration of resources and the limited range of the agents' movements and vision.

Figure 1 shows the population trajectories generated by the ABM with specified infertilities; and the continuous Verhuslt and discrete Hutchinson-Wright (Eq. 1) trajectories with appropriate intrinsic growth rates and time delay. The regimes of these trajectories move from stable on the right, to steady oscillations in the middle, to fully chaotic on the left based on increasing growth rates.

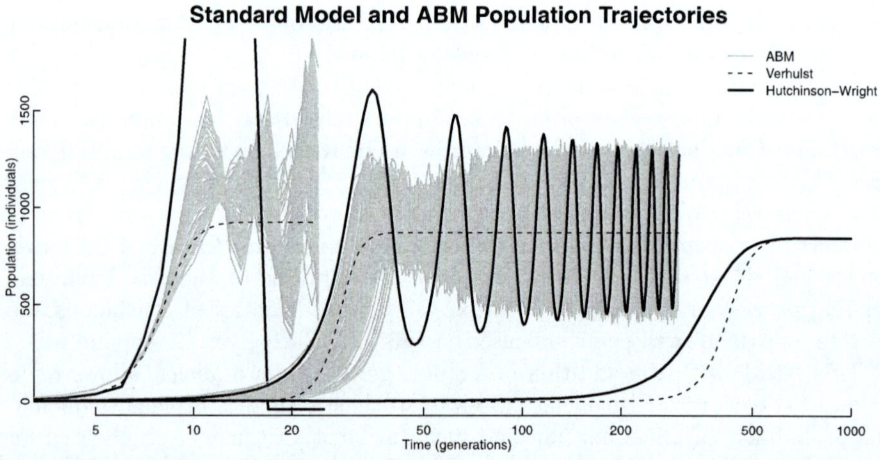

Fig. 1. Population trajectories. ABM population trajectories for infertility 1, 5 and 85 (from left to right) with best fit Verhulst intrinsic growth rate (Eq. 17) and Hutchinson-Wright (Eq. 1) intrinsic growth rate and delay coefficient.

These single species population models assume a well mixed, spatially homogeneous populations. Furthermore, only the number of individuals in the population over time is modeled. The replicating ABM also provides both spatial resolution and addition descriptors of the populations. These measures provide significant additional insights into the limits of the standard models and will be addressed in turn.

2.1 Relaxation of Specifications - Equilibrium

Equilibrium in these population models is considered to occur when the population has reached a steady size in the stable growth regime of the standard model. The ABM has additional measures of the population available when replicating these cases that reveal an underlying lack of equilibrium as measured by, for example, mean population age. Figure 2 demonstrates this mismatch in equilibrium times for both low growth and high growth (but not oscillatory) configurations.

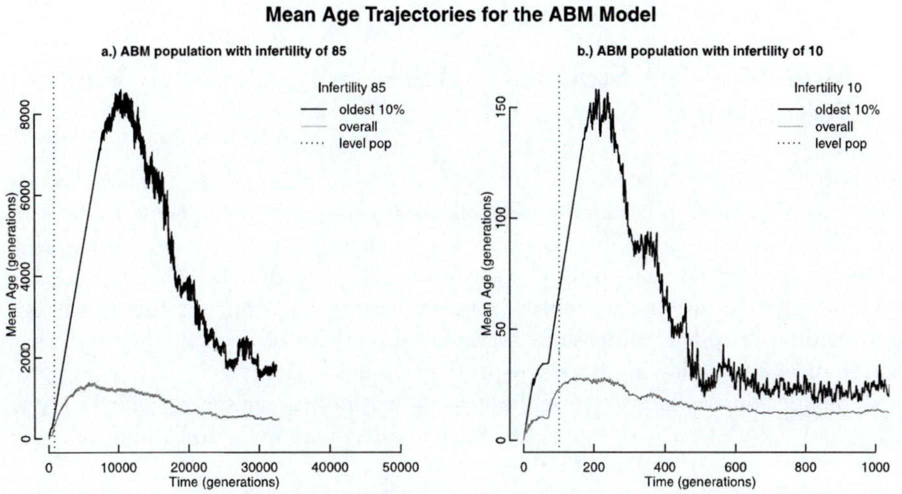

Fig. 2. Mean age equilibrium. ABM mean age trajectories for 2a) low growth (infertility of 85) and 2b) high growth (infertility of 10). Vertical dotted lines show the generation at which the population size reaches equilibrium.

These additional characteristics do not achieve equilibrium, sometimes for orders of magnitude greater than for the population size to reach equilibrium. The impact of this dynamic is significant for results based on these underlying metrics, for example wealth inequality [14–17,23] and allele diffusion (Sect. 3).

2.2 Relaxation of Specifications - Extinction

While the discrete logistic equation does generate extinctions, these extinctions occur for $N_{t+1} < 0$ and are due to exceptionally large values of N_t as shown by the Hutchinson-Wright trajectory at the far left of Fig. 1. The logistic equation is often modified (Ricker Curve) to avoid these "unrealistic" excursions [2]. The ABM with its spatial modeling as opposed to the well mixed specification of the standard model does not immediately go extinct for these cases due to spatial limits on reproduction.

In regimes of large oscillations, the replicating ABM generates different extinctions based on a lack of resources locally even though the global resources are more than adequate. These extinctions are driven by spatial waves of agents who consume everything in their path until they find themselves in a local desert of resources, unable to see or move to the ample food. Whether the population becomes extinct is dependent on a stochastic process where a few individuals may survive behind the wave to seed the next generation. With such a dependency on a stochastic process, extinction is inevitable. Figure 3 shows both events for the ABM: the top row where a few individuals survive behind the wave to seed the next generation and the bottom row where none survive and extinction occurs in a local desert.

3 Models of the Stochastic Behavior of Genes in Finite Populations

An important generalization of the Wright-Fisher class of models was introduced by Cannings [21]. This model allows overlapping generations as seen in the ABM simulations of populations. Its stochastic rule is based on the distribution of descendant genes. The second model considered, which also allows overlapping generations, was introduced by Moran [22] and has the advantage of providing "explicit expressions for many quantities of evolutionary interest" and is reversible, allowing for retrospective studies [4,5,11,12].

The specific configuration of these standard models chosen for this study is a haploid population with one gene and two alleles, no mutations, and, initially no selective pressure. The standard models assume a constant population, a specification which will be relaxed. For the Cannings model, the population at each $t+1$ generation is based on the random possibility of reproduction, and survival or death of each individual, over the entire population. For the Moran model, at generation $t+1$, one individual is selected for reproduction and one individual is then randomly selected to die, excluding only the newborn.

The population characteristics measured for this study are the probability of allele fixation, the mean time to absorption of a single allele in an otherwise homogeneous population of the other allele (of interest for survival of a single mutation), and the mean time to absorption for initial equal number of alleles. The specification of constant population and homogeneous mixing are then relaxed and the results discussed.

For all the Wright-Fisher class of models with no mutations or selective pressure, the probability π of one of the two alleles (A_1, A_2) overtaking the whole population (fixation) is given as:

$$\pi = i/2N \qquad (2)$$

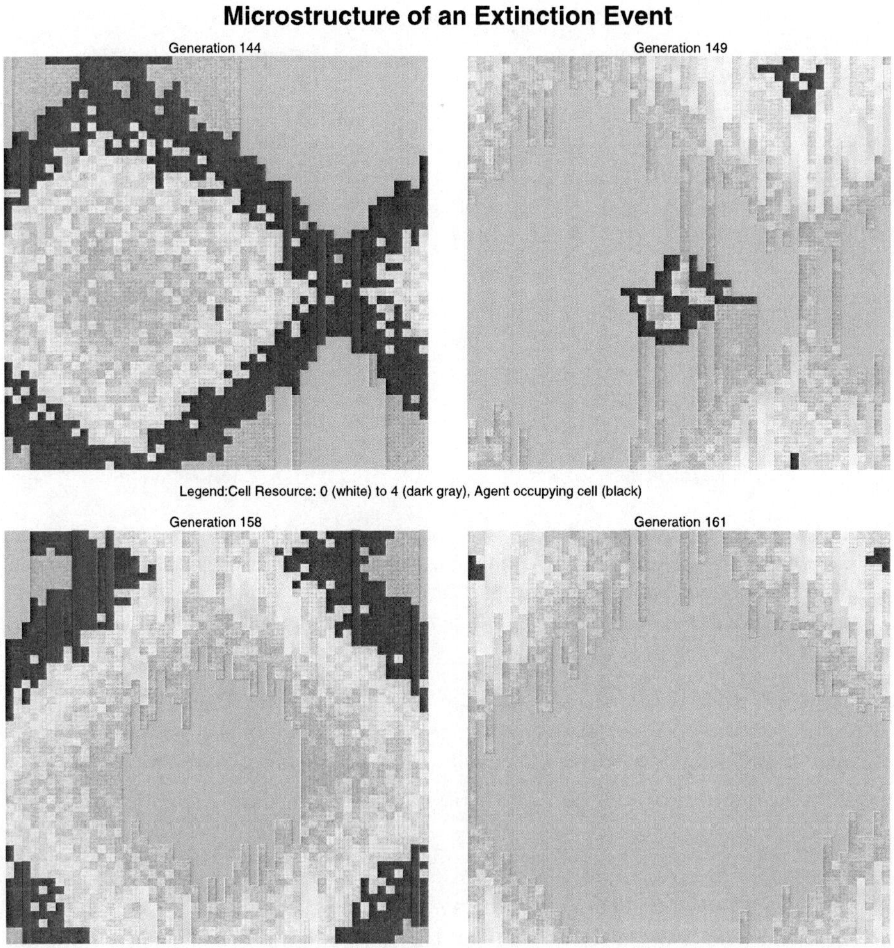

Fig. 3. Spatial extinction. a) When the penultimate wave of agents pass through the rich landscape, a few agents manage to survive behind the wave (Generation 144). As the last agents of the wave perish in a local desert, the next generation blooms (Generation 149). For the final wave, no agents survive behind the wave as it passes (Generation 158). The last cycle finds the remaining agents all in a local desert and about to perish (Generation 161). This extinction event is driven by non-equilibrium movement dynamics, and spatially local resource densities. Together these factors produce a highly stochastic process leading to certain extinction but at an unpredictable time.

where i is the initial population of one of the two alleles and the total population is $2N$. The selection of $2N$ for the haploid population allows comparison with diploid populations of standard Wright-Fischer models. Figure 4a shows this simple linear relationship as a function of i and the replication of these probabilities across the starting allele fraction for the ABM.

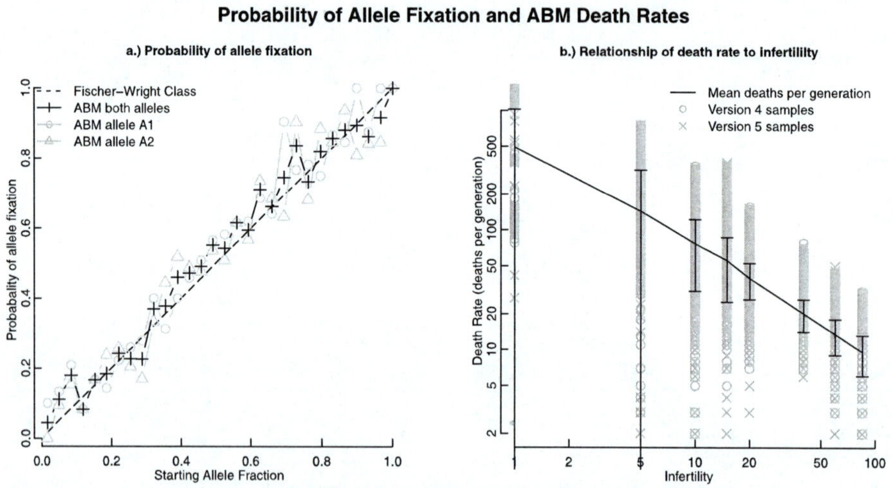

Fig. 4. Allele fixation and death rates. a) Comparison of ABM generated allele fixation probabilities versus allele starting frequency with the Wright-Fischer class of stochastic gene diffusion models. b) Relationship of ABM infertility to death rates to support the Cannings and Moran models.

Figure 4b demonstrates the one-to-one relationship of the ABM infertility parameter with the ABM-generated death rates. This relationship is used to map the replicating ABM to the Cannings and Moran models. The Cannings rule for stochastic reproduction of an individual a_i at time t can be expressed as vector $y(t)_i$

$$y(t)_i = [2, 2, 1, 0, 0, 1, ...0] \tag{3}$$

where y_i ranges across all the individuals a_i from $i = 1$ to $i = 2N$. The individual entries are the net population result for the descendants of each individual for that generation. Only one birth per individual per cycle is allowed. The variance, $\sigma(t)^2$, of $y(t)_i$ is

$$\sigma(t)^2 = \frac{1}{2N-1}\sum_{i=1}^{2N}[y(t)_i - \bar{y}(t)]^2 \tag{4}$$

where $\bar{y}(t)$ is mean of $y(t)_i$. With a constant sized population $\bar{y}(t) = 1$, the births equal the deaths $d(t)$, and (4) becomes

$$\sigma(t)^2 = \frac{1}{2N-1}[\sum_{1}^{d(t)}(2-1)^2 + \sum_{1}^{2N-2d(t)}(1-1)^2 + \sum_{1}^{d(t)}(0-1)^2] \tag{5}$$

where the number of individuals that survived but did not reproduce is equal to the total population minus those that gave birth minus those that died ($2N - 2d$). Simplifying

$$\sigma(t)^2 = \frac{1}{2N-1}[d(t) + 0 + d(t)] = \frac{2d(t)}{2N-1} \tag{6}$$

The ability to calculate this variance directly for the ABM is quite useful. With this variance, the mean time to allele absorption $\bar{t}_{abs}$ is given as [4]

$$\bar{t}_{abs} = -(4N-2)[p\log(p) + (1-p)\log(1-p)]/\sigma(t)^2 \quad (7)$$

where p is the ratio of the number of one allele, say A_1, to the total population.

This general Cannings result can also generate the mean absorption times for Wright-Fischer with $\sigma^2 \approx 1$ and Moran $\sigma^2 \approx 2/(2N)$. Note for Moran the generation t_m is a single individual's reproduction event and conversion to the Cannings t_c is written as

$$t_c = t_m/2N \quad (8)$$

The mean time to absorption for a single mutation A_1 in a population of A_2 is the mean time until the allele is lost, most likely, or takes over the population (unlikely but does occur with exceptionally long absorption times). Figure 5a gives the mean absorption times for a single allele, using the Cannings and Moran models from equation X and the results from the ABM runs. Figure 5b gives the results for these two standard models and the ABM over an even distribution of p.

3.1 Relaxation of Specification - Constant Population

As can be seen in Fig. 1, the population at steady state can only be considered constant for the dynamic regimes that are not oscillating nor chaotic. Surprisingly, as can be seen in Figs. 5a and 5b, the standard models are in general agreement with the ABM for all but the most extreme chaotic regimes. These results are also in agreement with recent, more complex, equation-based models of non-constant population [5] and spatial effects [12].

3.2 Weak Selection

Weak selection pressures are introduced to the ABM by modifying the agents' rule for movement towards the largest and the closest resources within range. If there are two different cells within range with the same largest resources and the same distance, the neutral ABM randomly selects between the ties. For the weak selection ABM, these ties are resolved based on one of the two alleles the agent carries: an introvert allele selects the target cell that has fewer other agents around it, and an extrovert allele that selects the target cell that has more agents surrounding it.

Using the Moran model [4], the probability that the next individual selected to die is an A_1 allele is

$$\mu_1 i/\{\mu_1 + \mu_2(2N-i)\} \quad (9)$$

where i is the number of A_1 alleles in the $2N$ population and $\frac{\mu_1}{\mu_2}$ defines the selective advantage. If $\mu_1 = \mu_2$ there is no selective advantage whereas if $\mu_1 < \mu_2$ then allele A_1 has a small selective advantage. The probability of fixation π is now:

$$\pi_i = \{1 - (\mu_1/\mu_2)^i\}/\{1 - (\mu_1/\mu_2)^{2N}\} \quad (10)$$

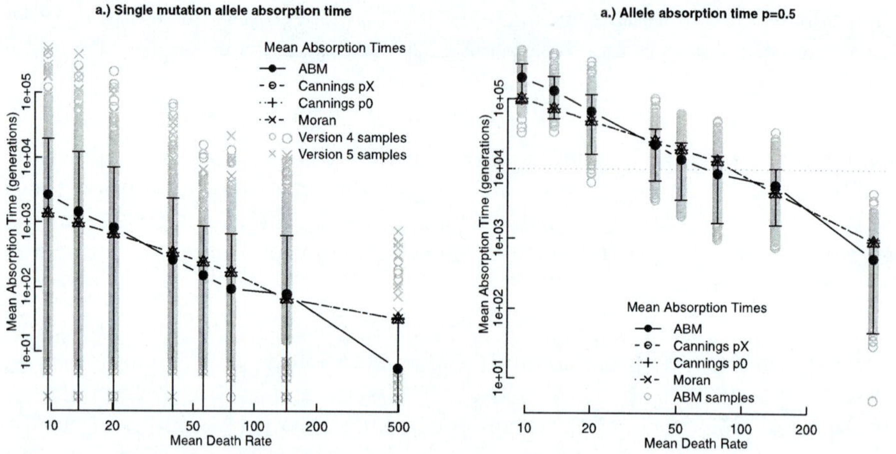

Fig. 5. Mean allele absorption times. a) The ABM, Cannings, and Moran mean absorption times for a single mutation allele. b) The allele absorption times for ABM, Cannings, and Moran models for an allele ration of 1/2.

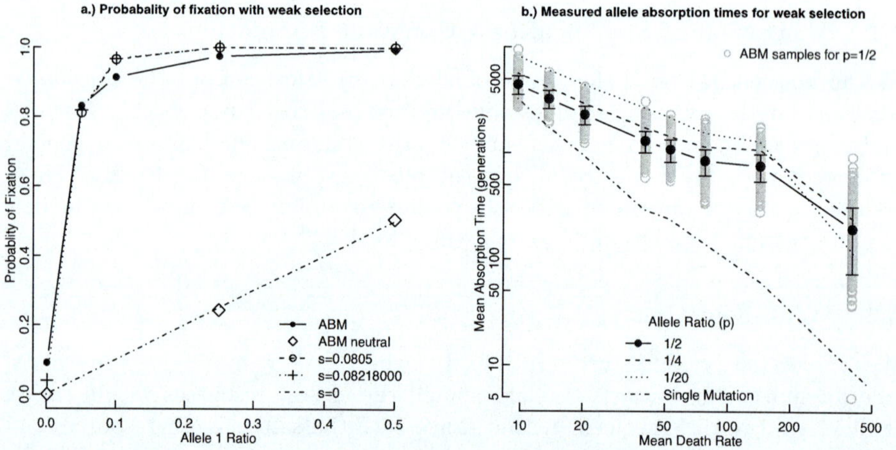

Fig. 6. Weak selection fixation and absorption. a) The probability of fixation of A_1. b) The mean time to absorption versus mean death rate for an allele ratio of 1/2.

By defining $\mu_1/\mu_2 = 1 - s/2$ with s small and positive, Eq. (10) can be approximated as

$$\pi(x) = \{1 - \exp(-\alpha x/2)\}/\{1 - \exp(-\alpha/2)\} \tag{11}$$

where $x = i/2N$ and $\alpha = 2Ns$. With this approximation, the magnitude of the selection pressure as represented by s can be implied from the ABM as on the order of 0.08, a somewhat weak selection pressure. Figure 6a presents the probability of fixation of the introvert and neutral phenotypes as a function of its initial ratio and the Moran standard model predictions of two of the implied parameters for weak and neutral selection pressure. Figure 6b presents the dramatically reduced absorption times. The single mutation fixation and absorption times diverge from the standard models predictions due to the stochastic effects on a single gene overriding the weak selection pressure, an expected divergence [4,5].

4 Conclusions

The ABM reproduced the results of the standard models well in the regimes that matched their specifications and foreshadowed recent equation based modelling research when the specifications were relaxed. For the Hutchinson-Wright time-delayed logistic model of population growth, no indication is available for when the population reaches a steady state since the underlying characteristics such as mean age may take many orders of magnitude more time than the population size to reach equilibrium. Not surprisingly the assumption of a well mixed population without a spatial dimension over predicts extinctions in high growth regimes and fails to predict extinctions due to local (spatial) resource depletion. The discrete stochastic models of gene frequency with and without selection, on the other hand, survived well into the oscillating population size regimes and only differed from the ABM at the highest chaotic regimes.

Appendix A - Computational Model and Process

Table 1 provides the definition of the agents' and landscape's parameters used for this study. Vision and movement are along rows and columns only. The two dimensional landscape wraps around the edges (often likened to a torus). Agents are selected for action in random order each cycle. The selected agent moves to the closest visible cell with the most resources with ties resolved randomly. After movement, the agent harvests and consumes (metabolizes) the required resources. At this point, if the agent's resources are depleted, the agent is removed from the landscape. Otherwise an agent of sufficient age (not a newborn) then considers reproduction, requiring a lucky roll of the fertility die (infertility), and an empty von Neumann neighbor cell, which are only the four neighboring spaces one step away by row or column. The newborn is placed in the empty cell and remains inactive until the next action cycle. With this approach for the action cycle, no endowments are required whether for new births or for the agent(s) at start-up. Once all the agents have cycled through, the landscape replenishes at the growth rate and the cycle ends.

Table 1. Agent and landscape parameters of the ABM

Agent characteristic	Notation	Value	Units	Purpose
Vision	v	6	Cells	Vision of resources on landscape
Movement	–	6	Cells per cycle	Movement about landscape
Metabolism	m	3	Resources per cycle	Consumption of resource
Birth cost	bc	0	Resources	Sunk cost for reproduction
Infertility	f	1–85	1/probability	Likelihood of birth
Puberty	p	1	Cycles	Age to start reproduction
Surplus	S	0+	Resources	Storage of resource across cycles

Landscape characteristic	Notation	Value	Units
Rows	–	50	Cells
Columns	–	50	Cells
Max capacity	R	4	Resource per cell
Growth	g	1	Resource per cycle per cell
Initial	R_0	4	Resource, all cells

The metabolism rate, uniform across a given population, is one that consumes per cycle 25% less than the maximum capacity per cell.

Control Volume Analysis

Figure 7 defines a control volume to track the flow of resources from generation on the landscape to destruction by the agents through metabolism, death, and birth. Note with no endowments, the only source of resources is growth on the landscape. Equations external to the ABM model are written below that provide verification, quantitative descriptions for replication, and analytic tools which reveal the internal workings of the ABM.

The change in resources within the landscape $\Delta E_L(t)$ per cycle can be written as

$$\Delta E_L(t) = \sum_{c=1}^{N_c} g_c(t-1) - H(t) \tag{12}$$

at the start of generation t, where N_c are the landscape cells, and the growth rate for each of the cells in the landscape $g_c(t)$ is given as:

$$g_c(t) = \begin{cases} g & r[c,t] + g \leq R \\ R - r[c,t] & r[c,t] + g > R \end{cases} \tag{13}$$

with R the maximum resources per cell, g the resource growth rate per cell, and $r[c,t]$ is the level of resources in the cell at t. The harvest of resources per cycle $H(t)$ is given as:

$$H(t) = \sum_{a=1}^{A(t-1)} r[c(a),t] + \delta_{p_0}(a) \sum_{a=1}^{B(t)} r[c(a),t] \tag{14}$$

where $c(a)$ is the cell location of agent a at the end of its action cycle, $A(t)$ are the agents alive at the end of cycle t, $B(t)$ are the agents born in cycle t,

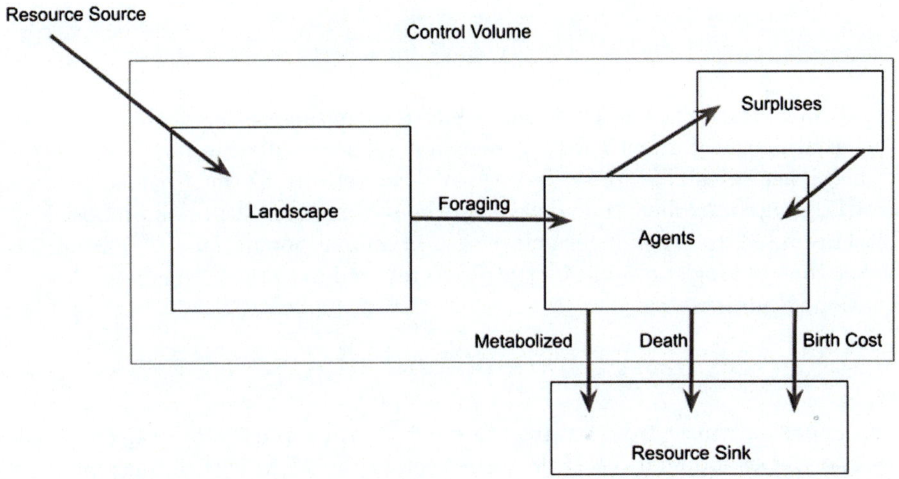

Fig. 7. Control volume. The flow of resources from the source in the landscape to the sinks of agents' consumption.

and $\delta_{p0}(a)$ is 1 if the duration of puberty is 0 and 0 otherwise. The change of resources within the agent population $\Delta E_p(t)$ is written as

$$\Delta E_P(t) = H(t) - \sum_{a=1}^{A(t)} m(a) - \sum_{a=1}^{D(t)} [S_a(t) + m(a)] - \sum_{a=1}^{B(t)} [bc(a) - \delta_{p_0}(a)m(a)] \quad (15)$$

where $m(a)$ and $bc(a)$ are the metabolism and birth cost of agent a respectively, $D(t)$ are the agents who died in this cycle and S_a is the surplus resources agent a brought into this cycle. By requiring a conservation of resources, equation for the balance of resources is given as:

$$0 = \Delta E_L(t) + \Delta E_P(t) + \sum_{a=1}^{A(t)} [S_a(t) - S_a(t-1)] \quad (16)$$

Appendix B - Single Species Models from Mathematical Biology

A continuous homogeneous model of a single species population $N(t)$ was proposed by Verhulst in 1838 [2]:

$$\frac{dN(t)}{dt} = rN(1 - \frac{N}{K}) \quad (17)$$

where K is the steady state carry capacity, t is time, and r is the intrinsic rate of growth. This model represents self-limiting, logistic growth of the population. This macroscopic model of a continuous, homogeneous population is quite descriptive and allows the exact solution

$$N(t) = \frac{K}{[1 + (\frac{K}{N_0} - 1)e^{-rt}]} \tag{18}$$

where N_0 is the initial population. While the continuous Verhulst Model fits the initial phase of growth well, it does not model oscillating population sizes at the higher rates of intrinsic growth. A discrete form of the Verhulst process incorporating an explicit time delay τ in the self-limiting term was proposed by Hutchinson [20] to account for delays seen in animal populations. The resulting discrete-delayed logistic equation [19], often referred to as the Hutchinson-Wright equation [3] is then

$$N(t+1) = [1 + r - \frac{N(t-\tau)}{K}]N(t) \tag{19}$$

This model's intrinsic growth rate with $\tau = 5$ captures the steady state, oscillating, and chaotic populations trajectories seen in the ABM with similar intrinsic growth rates.

References

1. Epstein, J.M., Axtell, R.: Growing Artificial Societies: Social Science from the Bottom Up. Brookings Institution Press (1996)
2. Murray, J.D.: Mathematical Biology. Springer, Heidelberg (2002)
3. Kot, M.: Elements of Mathematical Ecology. Cambridge University Press, Cambridge (2001)
4. Ewens, W.J.: Mathematical Population Genetics. Springer, New York (2004). https://doi.org/10.1007/978-0-387-21822-9
5. Parsons, T.L., Quince, C., Plotkin, J.P.: Absorption and fixation times for neutral and quasi-neutral populations with density dependence. Theor. Popul. Biol. **74**, 302–310 (2008)
6. Verhulst, P.F.: Notice sur la loi que la populations suit dans son accroissement. Corr. Math. Phys. **10**, 113–121 (1838)
7. Gause, G.F.: The Struggle for Existence. Williams and Wilkins (1934)
8. Binny, R.N., Law, R., Plank, M.J.: Living in groups: spatial-moment dynamics with neighbor-based movements. Ecol. Model. **415**(2020), 108825 (2020)
9. Bolker, B., Pacala, S.W.: Using moment equations to understand scholastically driven spatial pattern formation in ecological systems. Theor. Popul. Biol. **52**, 179–197 (1997)
10. Bordj, N., El Saadi, N.: Moment approximation of individual-based modes. Application to the study of the spatial dynamics of phytoplankton populations. Appl. Math. Comput. **412**, 126594 (2022)
11. Hathcock, D., Strogatz, S.H.: Fitness dependence of the fixation-time distributions for evolutionary dynamics on graphs. Phys. Rev. E **100**, 012408 (2019)
12. Tkadlec, J., Pavlogiannis, A., Chatterjee, K., Nowak, M.: Commun. Biol. **2**, 138 (2019)
13. Ovaskainen, O., Somervuo, P., Finkelshtein, D.: Mathematical and simulation methods for deriving the extinction thresholds in spatial and stochastic models of interacting agents. Methods Ecol. Evol. **12**, 162–169 (2021)
14. Adamou, A., Peters, O.: Dynamics of inequality. Significance **13**(3), 32–35 (2016)

15. Benisty, H.: Simple wealth distribution model causing inequality-induced crisis without external shocks. Phys. Rev. E **95**, 052307 (2017)
16. Berman, Y., Peters, O., Adamou, A.: Wealth inequality and the ergodic hypothesis: evidence from the United States. Working Paper (2020). https://ssrn.com/abstract=2794830
17. Benhabib, J., Bisin, A., Zhu, S.: The distribution of wealth and fiscal policy in economies with finitely lived agents. Econometrica **79**(1), 123–157 (2011)
18. Roughgarden, J., Bergmen, A., Hafir, S., Taylor, C.: Adaptive Computation in Ecology and Evolution. SFI Studies in the Sciences of Complexity, Vol. XXVI (1996)
19. Wright, E.M.: A non-linear difference-differential equation. J. für die reine angewandte Math. **194**, 66–87 (1955). https://doi.org/10.1515/crll.1955.194.66
20. Hutchinson, G.E.: Circular causal systems in ecology. Ann. New York Acad. Sci. **50**, 221–248 (1948). https://doi.org/10.1111/j.1749-6632
21. Cannings, C.: The latent roots of certain Markov chains arising in genetics: a new approach 1. Haploid models. Adv. Appl. Prob. **6**, 260–290 (1974)
22. Moran, P.A.P.: Random processes in genetics. Proc. Camb. Philos. Soc. **54**, 60–71 (1958)
23. Stevenson, J.C.: Dynamics of wealth inequality in simple artificial societies. In: 16th Annual Social Simulation Conference, September 2021
24. Liz, E.: Delayed logistic population models revisited. Proceedings of New Trends in Dynamical Systems, Salou (2012). https://doi.org/10.5565/PUBLMAT_Extra14_17

Correction to: Machine Learning Reveals Adaptive COVID-19 Narratives in Online Anti-Vaccination Network

Richard Sear, Rhys Leahy , Nicholas Johnson Restrepo ,
and Neil Johnson

Correction to:
Chapter "Machine Learning Reveals Adaptive COVID-19 Narratives in Online Anti-Vaccination Network" in:
Z. Yang and E. von Briesen (Eds.): ***Proceedings of the 2021 Conference of The Computational Social Science Society of the Americas*****, SPCOM,**
https://doi.org/10.1007/978-3-030-96188-6_12

In the original version of the chapter, the author "Yonatan Lupu" were included erroneously as co-author, which has now been corrected. The chapter and book have been updated with the changes.

The updated original version of this chapter can be found at
https://doi.org/10.1007/978-3-030-96188-6_12

Z. Yang and E. von Briesen (Eds.): CSSSA 2021, SPCOM, p. C1, 2023.
https://doi.org/10.1007/978-3-030-96188-6_14

Author Index

Z. Yang and E. von Briesen (Eds.): CSSSA 2021, SPCOM, p. 191, 2022.
https://doi.org/10.1007/978-3-030-96188-6